CENTRIFUGAL
PUMP DESIGN

CENTRIFUGAL PUMP DESIGN

John Tuzson

Evanston, Illinois

A WILEY-INTERSCIENCE PUBLICATION

JOHN WILEY & SONS, INC.

New York / Chichester / Weinheim / Brisbane / Singapore / Toronto

This book is printed on acid-free paper.

Copyright © 2000 by John Wiley & Sons, Inc. All rights reserved.

Published simultaneously in Canada.

No part of this publication may be reproduced, stored in a retrieval system or transmitted in any form or by any means, electronic, mechanical, photocopying, recording, scanning or otherwise, except as permitted under Sections 107 or 108 of the 1976 United States Copyright Act, without either the prior written permission of the Publisher, or authorization through payment of the appropriate per-copy fee to the Copyright Clearance Center, 222 Rosewood Drive, Danvers, MA 01923, (978) 750-8400, fax (978) 750-4744. Requests to the Publisher for permission should be addressed to the Permissions Department, John Wiley & Sons, Inc., 605 Third Avenue, New York, NY 10158-0012, (212) 850-6011, fax (212) 850-6008, E-Mail: PERMREQ @ WILEY.COM.

This publication is designed to provide accurate and authoritative information in regard to the subject matter covered. It is sold with the understanding that the publisher is not engaged in rendering professional services. If professional advice or other expert assistance is required, the services of a competent professional person should be sought.

Library of Congress Cataloging-in-Publication Data

Tuzson, John
   Centrifugal pump design / by John Tuzson.
     p. cm
   "A Wiley-Interscience publication."
   ISBN 0-471-36100-3 (cloth : alk. paper)
     1. Centrifugal pumps—Design and construction. I. Title.

  TJ919. T89 2000
  621.6'7 dc21   00-023096

Printed in the United States of America.

10 9 8 7 6 5 4 3 2 1

CONTENTS

Preface   xi

1. INTRODUCTION   1

Scope of book / 1
Information sources / 3
References / 3

2. FLUID MECHANIC FUNDAMENTALS   5

Practical numerical calculations / 5
Physical principles / 9
  Continuity / 9
  Balance of forces / 10
  Conservation of energy / 10
  Potential flow, rotational flow and shear / 12
  Curved and vortex flow / 13
  Coordinates relative to rotating axes / 14
References / 15

3. INSTABILITY, TURBULENCE AND SEPARATION   17

Reynolds number / 17
Instability and turbulence / 18
  Vortex flow / 20
  Separation / 20
  Lift / 23
  Drag / 24
References / 25

4. PIPES, BENDS AND DIFFUSERS   27

Pipes and bends / 27
Diffusers / 30
   Obstructions in pipes / 34
References / 34

5. PUMPING SYSTEMS   37

Pump specifications and selection / 37
System modeling / 41
References / 49

6. GENERAL PUMP THEORY   51

Pump performance / 51
Off-design operation / 53
Pump types and specific speed / 53
Pump configurations / 53
References / 60

7. PUMP FLUID MECHANICS   61

Impeller work input / 61
Flow at impeller inlet and exit / 63
Slip / 64
Relative velocity in the impeller and rothalpy / 67
Flow separation in the impeller / 69
Jet-wake flow pattern / 73
Blade-wake mixing / 75
Vaned diffuser / 75
Volute / 77
Incidence / 78
Cavitation / 80
Disk friction on the impeller / 82
Axial thrust on the impeller / 84
Leakage / 86
References / 87

8. BOUNDARY LAYERS IN PUMPS   91

Basic concept / 91
Turbulence / 92
Boundary layer thickness / 94
Boundary layers in closed flow passages / 96

Skin friction / 98
Boundary layers on rotating disks / 98
Boundary layers in vaneless diffusers / 100
Boundary layers inside rotating impellers / 102
Practical pump design calculations / 104
References / 105

9. CAVITATION IN PUMPS   107

Description / 107
Net positive suction head / 108
Cavitation at design conditions / 110
Cavitation at off-design conditions / 113
  Example / 114
Cavitation damage correlation / 116
Inducers / 116
Vibration / 117
References / 118

10. PERFORMANCE CALCULATION   121

Calculation procedure / 121
Impeller work input / 123
Flow losses / 124
Calibration on test data / 128
Results / 131
References / 131

11. PUMP DESIGN PROCEDURE   133

Published design procedures / 133
Choice of impeller diameter, exit width, and exit blade angle / 133
Impeller inlet diameter / 138
Inlet blade angle / 139
Vaned diffuser and crossover inlet / 142
Volute design / 144
Mechanical design / 146
Impeller blade design / 147
Manufacturing requirements / 149
Blade coordinates / 150
Computer calculation of blade coordinates / 153
References / 159

12. DESIGN EXAMPLES   161

Preliminary design and performance estimate / 161
Inlet velocity calculation / 166
Detailed three-dimensional blade design / 167
Two-dimensional blade design / 173
Return passage design / 175
Volute design example / 177
Concluding remarks / 181

13. ROTOR DYNAMICS OF PUMPS   183

Fundamentals of shaft vibration / 183
Shaft and impeller supported on two bearings / 185
References / 188

14. INLET AND EXIT RECIRCULATION   189

Recirculation phenomena / 189
Inlet recirculation / 190
Exit recirculation / 194
References / 198

15. TWO-PHASE FLOW IN PUMPS   201

Two-phase fluid bulk properties / 201
Experimental pump performance data / 203
Calculation models / 207
Fully separated flow model / 209
Comparison with test data / 211
Fill–spill fluid coupling / 212
References / 214

16. HIGH-VISCOSITY PUMPS   217

Performance estimates at high viscosity / 217
Design for high viscosity / 219
References / 221

17. SAND AND SLURRY EROSION IN PUMPS   223

Applications and requirements / 223
Slurry erosion mechanisms / 224
Erosion rate estimates / 226
Typical erosion pattern in slurry pumps / 227

Slurry erosion testing / 230
References / 232

18. EJECTOR PUMP SYSTEMS 235

System description / 235
System characteristics / 236
Ejector / 237
Ejector specifications and selection / 239
Ejector system performance map / 240
Design example / 243
References / 245

19. EVALUATION ON NEW FLUID MACHINERY CONCEPTS 247

Development criteria / 247
Energy transfer / 248
Specific speed / 249
Fluid machinery types / 250
Fluid machinery components and systems / 253
  Seals / 253
  Bearings / 254
  Drives / 256
Cost / 257
References / 258

20. COMPUTER PROGRAMS FOR PUMPS 263

Computer calculations / 263
References / 265

APPENDIX 267

INDEX 295

PREFACE

The publication of this book was prompted by my desire to make advanced fluid mechanic fundamentals available to practicing industrial pump designers. Unfortunately, a profound rift exists between the academic or research community and those on the factory floor. The unintended consequence of peer review is that researchers are motivated by the approval of their colleagues instead of by the users of advanced technology, the product designers. The academic rule of "publish or perish" generates a flood of publications that are incomprehensible, and therefore useless, to those who could make profitable use of the technology. This lack of technology transfer represents a great loss to the economy, a waste of resources, and ultimately, a loss to all humankind.

Recent books on centrifugal pump fluid mechanics and design could be listed in two groups. One recalls valuable empirical industrial experience but lacks a comprehensive description of fundamental fluid mechanical phenomena. The other describes advanced laboratory experiments and analytical studies using a formidable mathematical apparatus. The pay scale and job description of most industrial engineers, employed in pump fabrication and application do not allow the hiring of master's-level engineers having a background in advanced fluid mechanics and partial differential equations. Product designers must be generalists, not specialists. The challenge before us is to close this gap by insisting on fundamental understanding, but using a mode of presentation and communication that is comprehensible to those with a minimal engineering education.

Here is the place to thank all those who had a part in the ultimate appearance of this book. Room does not suffice to thank individually my teachers in Budapest, Innsbruck, Paris, and at MIT in Cambridge, as well as my many colleagues in industrial R&D and in ASME who made me think and challenged me to make sense. However, I must make an exception by expressing my special gratitude to Paul Hermann and Steve Zakem for their many helpful comments and suggestions.

John Tuzson
(*Tougeon*, as in *treasure*, or *tou(t) jaune* in French)
Evanston, Illinois

1

INTRODUCTION

SCOPE OF BOOK

The intent of this book is to offer to those interested in pump design, a clear, detailed picture of the flow pattern in centrifugal pumps and a simple pump design procedure based on fundamental understanding, not on empirical formulas. There is the hope and expectation that readers will learn the skill to design pumps rationally without falling back on unexplained formulas. Teachers might find a practical outline for their program of instruction. Since most elements of computer programs can be calculated by hand, such hand calculations can serve as exercise problems, the results being known from the computer program printout.

In today's competitive world, pump designers are tempted to just quickly copy existing pumps, or fall back on empirical design formulas established by statistical surveys of existing pumps. Such guidelines can be programmed into computers and can produce pump designs with the push of a button. Great problems arise, however, when the designs do not perform as expected. Without a deeper understanding of the flow pattern in pumps, the reasons for malfunctioning are incomprehensible. A competitive market also demands pump designs outperforming the competition. Statistically average designs may be adequate but will not provide a competitive edge and will leave the manufacturer in a "me too," catch-up posture. Superior designs can come only from a detailed understanding of the flow pattern in pumps.

Today's pump designers need not one but a portfolio of calculation procedures and computer programs. Even at the high speed of present computers and anticipated even higher speeds of future computers, the execution of complex computer calculations requires time, effort, and corresponding expense. Usually, the major portion of the time is not consumed by the running of the program but by the time needed to collect and input the necessary data. Reduction, plotting, and interpretation of the calculation results takes additional time. The amount of time and expense spent on the design of a certain pump must be in proportion to the value of the pump. Simple, fast, approximate calculations may suffice for smaller, inexpensive pumps. Large, expensive custom-engineered pumps warrant complex calculations. It is estimated that a satisfactory three-dimensional flow-velocity calculation in a centrifugal impeller may take anywhere from 200,000 to 1 million computational mesh points (Gopalakrishnan et al. 1995, Kaupert et al. 1996). The design procedure presented here describes, at each design step, the calculation that needs to be performed and defines the results that are expected from the calculations. Simple personal computer (PC) programs and their program listings are offered here to execute the calculation steps. These will give fast, approximate results which may be adequate for simple pumps. However, if more precise results are desired and are warranted, more detailed, commercially available computer programs can be substituted for the simple ones presented here. Regardless of the type of computer program being used, the designer must know and understand what needs to be calculated, how the results will be used to improve pump design, and how far the accuracy of the computer program can be trusted. Computer programs are only tools. The creative talent and skill of the designer are still needed. The designer must judge the amount of effort warranted by the design of the particular pump under consideration, and therefore must select those design tools that are appropriate from the available portfolio of computational procedures.

Modern hydraulic design procedures of axial flow pumps, which are not treated here, are based on very extensive airfoil theory. These concepts have been developed during the early years of aviation and gas turbine development. Because of strategic importance, vast amounts of test data and empirical information were accumulated from airfoils and rows of airfoils, cascades, in wind-tunnel experiments. Excellent design procedures were developed using analytical approaches combined with empirical correlations. Specialized concepts, variables, and a specialized vocabulary (span, chord, camber, stagger, pitch, cascade, solidity, deviation angle, stall) describe these design procedures. Attempts have been made to graft these concepts onto centrifugal turbomachinery design procedures. Up to a first order of approximation, a similarity exists between axial flow patterns in axial flow machinery and radial flow in centrifugal machines. Mathematically, flow patterns around airfoils in axial flow can be mapped onto a radial, circular field. But because of the effect of the Coriolis acceleration, which is negligible in axial flow but very important in radial flow, or the phenomenon of the jet-wake flow pattern in centrifugal turbomachinery, for example, the similarity remains defective. Wislicenus pointed out this fact in 1941 (Wislicenus 1941). Pump design procedures, some of which predate most aircraft development, generated over the years distinct concepts, and a different vocabulary, which demand separate treatment.

INFORMATION SOURCES

An unprecedented wealth of useful information is stored and is freely available in the technical libraries (Cooper 1999, Engeda 1999). However, despite computer searches in the Engineering Index and Applied Mechanics Review, the specific information needed remains difficult to find, especially on practical design issues. In addition to the references quoted in appropriate chapters of this book, collections on pump technology have been published, first of all, by the great engineering societies: the Fluids Engineering Division of the American Society of Mechanical Engineers (ASME), the Institution of Mechanical Engineers in England, the Vereins Deutscher Ingenieure (VDI) in Germany, the Societé Hydraulique de France, the Japanese Society of Mechanical Engineers (JSME), and the International Association for Hydraulic Research (IAHR). Each has an archival journal containing technical articles, available from technical libraries, but special symposia volumes often need to be ordered directly from the associations. Research institutions such as the Electric Power Research Institute (EPRI), the Von Karman Institute (VKI), and the Pfleiderer Institute, have their own series of publications. The Societé Hydraulique de France reports on French hydraulic machinery research in special issues (*La Houille Blanche*). Testing and rating standards have been established and published by the International Standards Organization (ISO), the Hydraulic Institute, ASME, IAHR, the American Petroleum Institute (API), and the VDI. Several universities have centers specializing in turbomachinery: Michigan State University, Pennsylvania State University, École Polytechnique Fédéral de Lausanne, and Braunschweig and Karlsruhe Universities. The Turbomachinery Laboratory of Texas A & M University has hosted 15 pump user conferences. Other conferences and refresher courses are offered by universities, trade associations, and private consultants. Manufacturers' trade associations of several European countries are represented by EUROPUMP, which supports the trade journal *World Pumps*, published by Elsevier in Oxford, a successor of the publication of the British Hydromechanic Research Association (BHRA), Cranfield, Bedford, England. Other trade journals, including *Pumps & Systems* and *Pumping Technology*, and trade journals in the chemical industry in the United States, contain practical information.

REFERENCES

Cooper, P. (1999): Perspective: The New Face of R&D—A Case Study of the Pump Industry, *ASME Journal of Fluids Engineering*, December, pp. 654–664. Also *Chemical Engineering*, February, pp. 84–88.

Engeda, A. (1999): From the Crystal Palace to the Pump Room, *Mechanical Engineering*, February, pp. 50–53. Also ASME Paper 98-GT-22.

Gopalakrishnan, S., Cugal, M., Ferman, R. (1995): Experimental and Theoretical Flow Field Analysis of Mixed Flow Pumps, *2nd International Conference on Pumps and Fans*, Tsinghua University, Beijing, October 17–20.

Kaupert, K. A., Holbein, P., Staubli, T. (1996): A First Analysis of Flow Field Hysteresis in a Pump Impeller, *ASME Journal of Fluids Engineering*, Vol. 118, pp. 685–691.

La Houille Blanche (1968): No. 2/3.

La Houille Blanche (1977): No. 7/8.

La Houille Blanche (1980): No. 1/2.

La Houille Blanche (1982): No. 2/3.

La Houille Blanche (1985): No. 5.

La Houille Blanche (1988): No. 7/8.

La Houille Blanche (1992): No. 7/8.

VDI Berichte, (1980): Nos. 371, (1981): No. 424.

VDI Forschungshefte (1968): Nos. 527 and 528, (1969): No. 535, (1970): No. 548

Wislicenus, G. F. (1941): Discussion, in C. A. Gongwer, A Theory of Cavitation Flow in Centrifugal-Pump Impeller, *ASME Transactions*, January, pp. 38–40.

2

FLUID MECHANIC FUNDAMENTALS

PRACTICAL NUMERICAL CALCULATIONS

The fluid mechanical phenomena in centrifugal pumps are described in terms of velocity and pressure distributions. Local velocities are measured relative to the flow passage walls or the objects placed in the flow. The flow pattern is the same whether an airplane flies through still air or the air flows past a stationary airplane.

The pressure represents the pressure energy of a unit volume of fluid, as can be seen from its dimensions:

$$\frac{\text{energy}}{\text{volume}} = (\text{in-lb})/\text{in}^3 = \text{lb}/\text{in}^2 = \text{psi} = \text{pounds per square inch}$$
$$= (\text{N}/\text{m}^2)/\text{m}^3 = \text{N}/\text{m}^2 = \text{Pa} = \text{pascal}$$

In the case of incompressible fluids such as water, the pressure appears only in the form of a pressure difference with respect to a reference pressure. In pumps the pressure level is often given with respect to the pressure prevailing at the pump inlet. However, the choice of reference pressure is arbitrary as long as it is clearly defined. In practical applications the reference pressure level is often taken at a free water surface exposed to the ambient. In pressure-measuring instruments, the pressure is shown either with respect to ambient pressure, which at standard conditions is taken as $14.7 \, \text{lb}/\text{in}^2$, or is measured deliberately as a pressure difference, by connecting a manometer or differential pressure gauge to two locations in the fluid. Barometers measure pressure level with respect to an absolute vacuum. Consequently, the units of pressure are designated psig if they are measured from ambient pressure, and psia if they represent absolute pressure.

Unlike the case of incompressible fluids, absolute pressures must be used when the flow of compressible fluids is analyzed. Compressible fluids are typically gases, but two-phase flow, liquid and gas mixtures, also qualify as compressible fluids. In pumps, compressible flow phenomena appear when the liquid contains gases or vapors. Such a case is when cavitation appears, when the local pressure is lowered to the point where the liquid starts boiling and vapor bubbles appear. In compressible fluids, in general, fluid properties, such as the boiling point at a given temperature, and in particular the specific volume, depend on the absolute pressure, not on the pressure difference. Cavitation will appear when the absolute pressure anywhere in a flow dips below the vapor pressure of the fluid, which is a material property of the fluid depending on the absolute pressure and temperature. In such cases, absolute pressures should preferably be used.

Pump fluid mechanic calculation will require algebraic manipulations of the fundamental equations to arrive at numerical results. Many errors of numerical calculations can be avoided by careful balancing of the units of physical quantities. Regardless of the type of units used, SI or English, the three dimensions of mass M, length L, and time T enter into the units of the physical quantities of interest in pump design (Table 2.1).

Every term of an equation, consisting of products of variables, must have the same combination of dimensions. Indeed, they must even have the same units. This requirement helps to keep track of conversion factors and to observe the relationship between the units of mass and of force. (Force is mass times acceleration.) Applied, for example, to *Bernoulli's equation*, which states that the

Table 2.1 Dimensions of variables

Quantity	Dimension
Length, l	L
Time, t	T
Mass, m	M
Force, F, weight	ML/T^2
Pressure, p	M/LT^2
Flow rate, Q	L^3/T
Velocity, V	L/T
Acceleration, a	L/T^2
Work, energy, E	ML^2/T^2
Power, P	ML^2/T^3
Density, ρ	M/L^3
Specific weight, ρg	M/L^2T^2
Kinematic viscosity, ν	L^2/T
Dynamic viscosity, μ	M/LT

energy of a fluid element remains the same as it moves along a streamline, we have

$$\rho \frac{V^2}{2} + p + \rho g h$$

$$= \rho(M/L^3)\frac{V^2(L^2/T^2)}{2} + p(M/LT^2) + \rho(M/L^3)g(L/T^2)h(L)$$

$$= \frac{\text{energy}}{\text{volume}}(M/LT^2)$$

In English units:

$$\frac{\rho(\text{lb-sec}^2/\text{ft}^4)V^2(\text{ft}^2/\text{sec}^2)}{2} + p(\text{lb/in}^2)144(\text{in}^2/\text{ft}^2) + \rho(\text{lb-sec}^2/\text{ft}^4)g(\text{ft/sec}^2)h(\text{ft})$$

$$= \frac{\text{energy}}{\text{unit volume}}(\text{ft-lb/ft}^3) = \text{pressure}$$

In SI units:

$$\rho(N \cdot s^2/m^4)\frac{V^2(m^2/s^2)}{2} + p(N/m^2) + \rho(N \cdot s^2/m^4)g(m/s^2)h(m)$$

$$= \frac{\text{energy}}{\text{unit volume}}(N/m/m^3) = \text{pressure}$$

The various terms of the Bernoulli equation represent different types of energies. The ratios of terms yield dimensionless numbers, the magnitude of which expresses the relative importance of these energies. Such dimensionless numbers are:

$$\text{pressure coefficient} = \frac{\text{pressure energy}}{\text{inertia}} = C_p = \frac{p}{\rho V^2/2}$$

where p is the pressure, V the velocity, and ρ the density.

$$\text{Froude number} = \frac{\text{inertia}}{\text{gravitational or potential energy}} = \text{Fr} = \frac{V^2}{gh}$$

where V is the velocity, g the gravitational acceleration, and h the typical height, the density being eliminated.

The Bernoulli equation assumes inviscid flow; therefore, the viscous energy is neglected and does not appear. Nonetheless, the importance of viscosity can be expressed as the ratio of the inertia (ρV^2) and viscous energy ($\mu V/D$), which becomes the Reynolds number $= VD/\nu$, since the kinematic viscosity ν is the ratio of the dynamic viscosity μ and the density ρ. Flow calculations can be simplified considerably if either the viscous or inertia forces can be neglected. The magnitude of the Reynolds number serves as a measure to decide which force can be neglected. In pumping machinery handling water, large Reynolds

8 FLUID MECHANIC FUNDAMENTALS

numbers, typically well above the transition Reynolds number range of 10^2 to 10^3, are usually encountered, which show that viscous forces are not very important. The prevalence of viscous or inviscid flow conditions in various pump components is governed by the local Reynolds number based on a typical velocity V and on the typical geometrical width or size of the component D.

The magnitude of the pressure coefficient indicates, for example, how much pressure head is recovered in diffusers, and therefore becomes a measure of performance. Since it is dimensionless, it applies to similar diffusers regardless of their size and of the actual magnitude of the velocity.

Nondimensionalizing the fundamental equations by dividing through with the typical variables of one of the terms—for example, the inertia term—eliminates all units and dimensions. The solution can then be obtained in pure numbers or dimensionless coefficients, which have general validity. Such numbers are, for example, pipe friction loss coefficients, which specify that fraction of the velocity head, $\rho V^2/2$, which results in a pressure loss, $p_2 - p_1$.

When exploring the relationship between a number of variables, a dimensionless correlation can be established by forming dimensionless groups of each of the variables with the help of selected, representative values of the three basic dimensions—length, time, and mass—or other, selected variables having a dimension that is a combination of these basic dimensions. The three basic dimensions suffice for describing mechanical or fluid-mechanic phenomena. Further fundamental units must be introduced when other physical phenomena need to be analyzed, such as heat or mass transfer, electricity, or chemical processes.

Since several possible combinations exist, several possible equivalent sets of dimensionless groups, including all variables, can exist. The unknown relationship must involve one complete set of these dimensionless groups, even if the exact nature of their relationship is not known. Sometimes a relationship consisting of a product of all dimensionless groups is assumed, and the exponents of the dimensionless groups are determined by experiment. In particular, heat and mass transfer calculations in chemical engineering often use such correlations. In fluid mechanics such an empirical correlation has been established for estimating the pipe friction pressure drop. The dimensionless groups are the loss or pressure coefficient C_p, the friction coefficient λ, the ratio of length to diameter L/D, the relative roughness k/D, and the Reynolds number Re.

$$C_p = f(L/D, k/D, \text{Re}) = \lambda(L/D)$$

where λ becomes a function of k/D, and Re. Indeed, in the laminar and turbulent range, respectively (Schlichting 1960, p. 525):

$$\lambda = \frac{64}{\text{Re}}$$

$$\lambda = \frac{1}{[2\log(0.5D/k) + 1.74]^2}$$

Between the laminar and turbulent ranges a transition range exists, which is a more complex function of k/D and Re and is shown graphically on the empirical plot of the pipe friction coefficient. Similar but slightly different mathematical expressions have been derived by various scientists.

PHYSICAL PRINCIPLES

The fundamental equations, which are presented here, are not of the most general form but have been simplified to be used for designing pumps. Still they derive from the general physical principles of conservation of mass and energy and the balance of forces and moments (Batchelor 1967, Comolet 1994, Lamb 1955). Drastic simplifications are made here by assuming an incompressible fluid, most often water, and assuming steady flow. The effects of viscosity are neglected and the flow is assumed to be inviscid. Some special effects of viscosity are discussed in appropriate chapters.

To derive the mathematical equations corresponding to the principles described above, a space or volume—a control volume—is defined within the fluid. The closed boundaries or surface of the volume can be drawn arbitrarily. If the volume is chosen infinitesimally small—for example, the cube dx, dy, dz in rectangular coordinates—the equations appear in the form of partial differential equations of the three velocity components and the pressure. The technical literature refers to these partial differential equations as the Navier–Stokes equations. If the size of the volume is chosen larger than infinitesimal, the velocities entering or leaving the volume and the pressure acting on the boundaries must be integrated around the boundaries. In practice, the integration can be implemented by subdividing the surface into smaller elements, over which the flow velocity or pressure remains constant, and by adding up flow across these elements and the pressure forces. In many cases the boundaries can be chosen in such a manner that the velocity remains uniform and perpendicular or parallel to at least a portion of the boundary, which simplifies the calculations and leads to simple algebraic equations.

The state of the fluid at any one location is defined by its velocity and pressure. Since the velocity is a vector quantity and has not only a magnitude but also a direction in three-dimensional space, components of the velocity in three perpendicular directions must be known. The total number of unknowns is then four, and therefore four equations are required to obtain a solution. In many cases the flow can be assumed to take place in a plane, in two dimensions, or along a line, in which case only three or two unknowns exist and only three or two equations are required. For example, the flow is often assumed to be parallel to the flow passage walls.

Continuity

The *equation of conservation of mass*, also called the *equation of continuity*, states that the steady flow into any space must be equal to the outflow. In steady

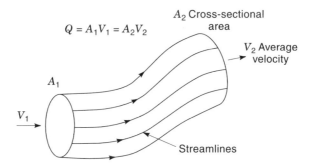

Figure 2.1 Continuity or conservation of mass equation.

flow the fluid particles follow certain streamlines. A flow cross section can be defined perpendicular to the limiting streamlines on the circumference of the cross section (Fig. 2.1). Then the flow through any cross section defined by the same streamlines must be the same. For example, an obvious choice of streamlines in a pipe corresponds to the pipe walls. The flow through any pipe cross section must then be the same.

$$\rho V_1 A_1 = \rho V_2 A_2$$

Balance of Forces

The three equations corresponding to the balance of forces must be written separately for the three perpendicular directions—x, y, z in rectangular coordinates—because they are independent from each other. The physical principle states that the pressure forces on the boundaries and the difference between the momentum entering and leaving the control volume in the chosen direction must be equal. The momentum of the flow is equal to the mass times velocity. In the simple case of straight, parallel flow, only one equation is needed. If there is an object in the flow, the force on the object, F, must be included in the equation (Fig. 2.2):

$$V_1(\rho V_1)A_1 + p_1 A_2 = V_2(\rho V_2)A_2 + p_2 A_2 + F$$

Conservation of Energy

The principle of *conservation of energy* states that the energy of a unit fluid element remains the same along the streamline it follows, provided that there are no losses. Since energy is a scalar quantity, only one equation is needed, the familiar Bernoulli's equation (Fig. 2.3):

$$p_0 = \frac{\rho V_1^2}{2} + p_1 + \rho g h_1 = \frac{\rho V_2^2}{2} + p_2 + \rho g h_2 + \text{losses}$$

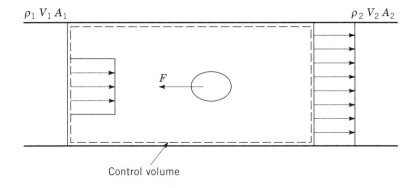

Figure 2.2 Balance of forces equation.

Here p_0 is the total pressure or stagnation pressure, the sum of all the terms. The variables h_1 and h_2 stand for the height of the two locations, measured from the same reference level, as shown on Fig. 2.3. When the equation is applied to short piping components or pumps, the head difference from the level difference is negligible compared with the head change due to head loss or head rise.

In pump calculations the terms of this equation are often divided through by the specific weight of the fluid, ρg, to convert to the terms of velocity head, pressure head, and static head all having the dimension of length. The total head or stagnation head, h_0, designates the sum of all these terms, and corresponds to

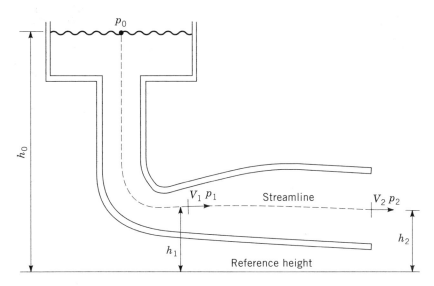

Figure 2.3 Conservation of energy equation.

12 FLUID MECHANIC FUNDAMENTALS

the head that would be reached if the fluid were brought to rest without any losses:

$$h_{01} = \frac{V_1^2}{2g} + \frac{p_1}{\rho g} + h_1 = \frac{V_2^2}{2g} + \frac{p_2}{\rho g} + h_2 + \text{losses}$$

The continuity equation and the three balance-of-forces equations are sufficient to determine the four unknowns: the pressure and the three velocity components. However, sometimes use of the energy equation is to be preferred to the force balance equation, especially if the flow is in one direction only. Also in some instances the losses can be calculated using the energy equation after the velocity and pressure have been calculated from the other equations. Such an example, that of the sudden expansion loss, will be given in Chapter 4.

Potential Flow, Rotational Flow, and Shear

If there are no losses and the energy—the total or stagnation pressure as expressed by Bernoulli's equation—is the same on all streamlines everywhere, then an irrotational, potential flow exists and calculation is considerably simplified (Flügge 1959, Lamb 1955). The important issue from the point of view of mathematics is whether the flow velocity vector field satisfies the conditions of irrotation. The important issue from the point of view of physics is whether the energy on all streamlines is the same. The two criteria are identical, as can be shown by mathematical deduction. (If the derivative of Bernoulli's equation is taken in the three coordinate directions, and if the corresponding Navier–Stokes equation, without viscous terms, is subtracted from each, it is found that the energy does not change in any direction, provided that the condition of irrotationality is satisfied.)

The irrotational, potential flow, or equal-energy hypothesis allows that the pressure be eliminated from the basic equations, which then contain only the velocities and the geometrical variables. All velocities can be changed by a common factor, positive or negative, without changing the shape of the streamlines. The flow can even reverse. The equations are linear and can be superposed. The sum of two or more solutions is also a solution. At every point, the velocities, calculated from different solutions, can be added. However, potential flow equations express only purely kinematic relationships. The flow can be unsteady and the calculated flow pattern can be an instantaneous snapshot of a time-dependent flow. The dynamics of the flow, the balance of forces, depends on the pressure distribution, which remains totally undefined until the pressure distribution on the boundaries has been specified. It can be reasoned that the stability of a potential flow solution, whether it will persist or change with time, also remains undefined until the dynamics of the flow is clarified. This consideration would imply that some potential flow solutions, which are unstable, cannot persist and will not be found in real life (D'Alambert's paradox, which states that objects in a hypothetical potential flow can have no drag, an obvious impossibility in real life).

Since the flow entering the pump can usually be assumed to have the same energy on all streamlines, a potential flow assumption is not unreasonable, provided that there are no major energy losses on some of the streamlines and provided that the flow follows the flow passage walls and does not separate. If parallel streamlines have different energies, a rotational flow exists with shear between neighboring streamlines. Shear flows can develop when the fluid on one streamline experiences greater energy losses than on another, or when pumps impart less energy to the fluid on one streamline, than on the other. In straight, parallel, rotational flow the velocity varies perpendicularly to the flow direction. The effects of shear are discussed in a subsequent chapter.

Curved Flow and Vortex Flow

If the streamlines are curved, the pressure increases perpendicular to the streamlines in the direction away from the center of curvature, because of the centrifugal acceleration. Conversely, if the streamlines are straight and parallel, the pressure must be the same throughout the cross section perpendicular to the streamlines. These rules help to visualize flow pattern in a given case and can provide some indication in which direction pressure and velocity change.

In axisymmetric, swirling flow, the use of polar coordinates—r, θ, z and in the radial, circumferential, and axial directions—is preferred. Axial symmetry exists and the flow variables do not depend on the circumferential angle θ. In the force balance equations in the three coordinate directions—the Navier–Stokes equations in cylindrical coordinates—two additional terms appear in addition to the terms corresponding to accelerations in the direction of the velocities, which result from the circumferential, rotational component of the velocity V_t. The inertia force due to the centrifugal acceleration $\rho V_t^2/r$ acts in the radial direction, and the inertia force due to the Coriolis acceleration $\rho V_t V_r/r$ acts tangentially in the direction opposed to the direction of the tangential velocity, provided that the radial velocity component V_r points in the direction of increasing radius, as shown in Fig. 2.4. These equations describe the flow with respect to a stationary frame of reference in terms of cylindrical coordinates (Batchelor 1967). The additional terms appear because of the curvature of the coordinate system. When applied to flow in rotors, such as pump impellers, clear distinction must be made between the absolute and relative variables, the absolute frame of reference being fixed, while the relative frame of reference rotates with the rotor.

Vortex flow represents a special case of curved streamlines, when the streamlines are concentric circles around the center of the vortex. The pressure increases with the radius: $dp/dr = \rho V^2/r$ because of the centrifugal acceleration. A potential vortex exists if the energy on all streamlines, everywhere, is the same. To arrive at an expression for the radial variation of the tangential velocity V, the pressure is eliminated from the energy equation, Bernoulli's equation, by

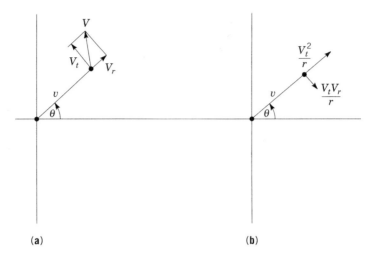

Figure 2.4 Axisymmetric flow in cylindrical coordinates: (a) velocities; (b) centrifugal and Coriolis accelerations.

differentiating it with respect to r and substituting dp/dr from the expression above.

$$\frac{d(p+\rho V^2/2)}{dr} = \frac{dp}{dr} + \rho V \frac{dV}{dr} = \frac{\rho V^2}{r} + \rho V \frac{dV}{dr} = 0$$

$$\frac{dV}{dr} = \frac{-V}{r} = \frac{d(C/r^2)}{dr}$$

$$Vr = C = \text{potential vortex}$$

Assuming that the axial velocity in the z direction is constant or zero, the solution of the resulting differential equation predicts that the tangential velocity V must vary in inverse proportion to the radius. Consequently, $Vr = $ constant.

The product of the circumferential velocity and the radius Vr corresponds to the angular momentum of the fluid particle. The finding above states that the angular momentum of a rotating fluid is conserved if there are no losses or external forces—a fundamental principle of physics. As will be seen, centrifugal pumps impart energy to the fluid by increasing its angular momentum. Consequently, this principle plays a very important role in pump calculations. The corresponding equations in the absolute as well as in the relative rotating frame of reference are given in Chapter 7 when the flow in a rotating pump impeller is described.

Coordinates Relative to Rotating Axes

Since flow in pump impellers can be considered axisymmetric, or shows an axial symmetry, polar coordinates are particularly convenient. Flow patterns relative to the rotating impeller can be studied by subtracting the rotational speed of the impeller, ωr, from the absolute tangential velocity. However, the additional terms corresponding to the centrifugal and Coriolis acceleration must be retained. The centrifugal and Coriolis accelerations distinguish the flow conditions in the impeller from the conditions that would be present if the impeller would not rotate. Although the flow pattern would be the same in a rotating and a stationary impeller if potential flow could be assumed, the pressure distribution would differ greatly. In the case of unequal energy flows, shear flows, when boundary layers or flow separation were present, secondary flows and unusual flow patterns can appear, such as a jet-wake flow or blade suction side separation, which would have no equivalent in an impeller that would not rotate.

If the flow is to be defined in terms of the relative velocities W_r, W_θ, and W_z with respect to the r, θ, and z coordinate directions rotating with an angular velocity ω around the z axis, the following expressions must be substituted for the absolute velocities into the equations in cylindrical coordinates: $V_r = W_r$, $V_\theta = W_\theta + \omega r$, and $V_z = W_z$. It is found that in addition to the terms one would see in the equations for a stationary, cylindrical coordinate system, the following acceleration terms appear in the radial and the circumferential directions which result from the rotation: $-2W_\theta \omega - \omega^2 r$ and $+2W_r \omega$. One recognizes the centrifugal acceleration term due to the rotational velocity $\omega^2 r$. The remaining two terms are due to components of the Coriolis acceleration expressed in the rotating frame of reference. These accelerations would not exist if the coordinate system would not rotate (Batchelor 1967).

REFERENCES

Batchelor, G. K. (1967): *Fluid Dynamics*, Cambridge University Press, New York.

Comolet, R. (1994): *Mécanique Expérimentale des Fluides*, Masson, Paris.

Flügge, S. (1959): *Encyclopedia of Physics*, Vol. VIII/1, Fluid Dynamics I, Springer Verlag, Berlin, pp. 187.

Lamb, H. (1955): *Hydrodynamics*, Dover, New York.

Schlichting, H. (1960): *Boundary Layer Theory*, McGraw-Hill, New York.

3

INSTABILITY, TURBULENCE, AND SEPARATION

REYNOLDS NUMBER

Drastic changes take place in the flow pattern when, with increasing velocity, the inertia forces become greater than the viscous forces. The ratio of these forces, the *Reynolds number*, gives a measure of their relative importance. In high Reynolds number flow, inertia forces predominate.

$$\text{Reynolds number} = \frac{\rho V^2}{\mu V/L} = \frac{\rho VL}{\mu} = \frac{VL}{\nu}$$

Here ρV^2 stands for the inertia forces and $\mu V/L$ for the viscous forces. The ratio of the dynamic viscosity μ and the density ρ is generally known as the kinematic viscosity ν.

The shear force due to the viscosity of the fluid acts between neighboring streamlines in the direction of the velocity. Its effect appears most clearly in the case of straight, parallel flow. In straight, parallel flow a viscous shear force τ appears when the velocity varies perpendicular to the streamlines (Fig. 3.1). It is proportional to the velocity gradient and perpendicular to the direction of the velocity dV/dy, the coefficient of proportionality being μ the dynamic viscosity of the fluid:

$$\tau = \mu \frac{dV}{dy}$$

18 INSTABILITY, TURBULENCE, AND SEPARATION

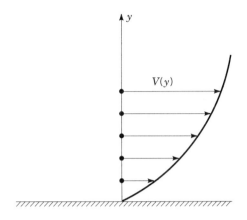

Figure 3.1 Parallel shear flow.

At 68 °F or 20 °C the numerical values of these quantities for water are $\mu = 21.1 \times 10^{-6}$ lbf-sec/ft^2 = 1.0087cP, $\nu = 10.9 \times 10^{-6}$ ft^2/sec = 1.008×10^{-6} m^2/s, and $\rho = 1.935$ lbf-sec^2/ft^4 = 1.00×10^{-3} N·s^2/m^4.

Such shear appears, for example, at the walls of flow passages. The fluid immediately next to the wall does not move, and the velocity parallel to the wall increases gradually with the distance from the wall toward the main stream. A small, low-velocity layer—a boundary layer—exists near the wall, the thickness of which depends on the Reynolds number, and is typically small in water pumps. Since this shear force at the wall slows the flow and results in an energy loss, it must be taken into account in flow calculations, which is done by introducing a wall friction loss and a dimensionless pressure coefficient, the skin friction coefficient. Shear layers also exist on the sides of jets, where the high velocity of the jet neighbors a stagnant fluid. Wake boundaries, behind objects in a free stream, are also shear layers (Batchelor 1967, Crighton 1985).

It will be noted that in the straight, parallel flow case above, the velocity will vary from one streamline to the next only if the energy on the streamlines is different. Indeed, in straight, parallel flow, in the absence of streamline curvature, the pressure cannot vary perpendicular to the streamlines. Therefore, the sum of the pressure head and velocity head, the total or stagnation head, cannot remain the same. This state implies that the flow is rotational and is not a potential constant-energy flow. More generally stated, shear, and therefore viscous shear forces, can exist only in rotational flows.

INSTABILITY AND TURBULENCE

When viscous forces become small compared with inertia forces, waves and fluctuations appear on the shear layers between neighboring streamlines with different energies (Fig. 3.2). At moderately high Reynolds numbers these waves

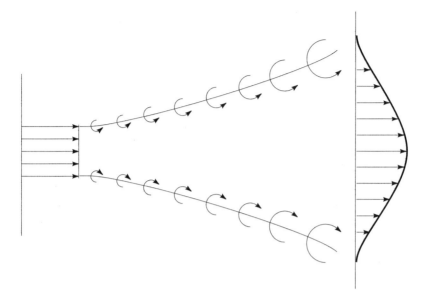

Figure 3.2 Unstable shear layer on jet boundary.

grow and eventually, farther downstream, roll up in vortices. At high Reynolds numbers an irregular fluctuation—turbulence—appears. This instability of shear flows is inherent in the nature of fluid flow and does not depend on viscosity. On the contrary, it appears when viscosity becomes unimportant. Turbulence is not a viscous phenomenon.

Turbulent fluctuations and excursions of fluid particles transfer energy from one streamline to the other. They accelerate the slow stream and slow the fast. Consequently, turbulence acts in a manner similar to viscosity. Therefore, an artificial eddy viscosity has been defined, which as a first approximation is sometimes used in calculations, as if it were the real viscosity of the fluid.

Practical flow calculations aim to arrive at the steady value of the flow variables. Calculation of the instantaneous value of the fluctuating quantities is of little interest. Therefore, advanced models to calculate turbulent flow operate with time-averaged values of the instantaneous velocities. These models attempt to represent the average effect of turbulent fluctuations. Several more advanced and more complex computational turbulence models have been developed. They still incorporate empirical data and usually give fully satisfactory results only for those kinds of flow configurations for which they have been developed.

Above it was stated that in straight, parallel flow, shear layers become unstable, develop waves, and roll up in vortices. Quite generally, in mechanical systems, the proof of stability consists of disturbing the equilibrium position and observing whether the system returns to it—whether the oscillations around the equilibrium position tend to damp out or amplify. Such calculations become prohibitively complicated in fluid mechanics. Nonetheless, it was possible to show that shear layers can be stabilized by an acceleration, and corresponding pressure gradient, oriented perpendicularly to the shear layer and in the direction in which the energy increases on neighboring streamlines (Bayly and Orszag 1988; Johnston et al. 1972; Tuzson 1977, 1993). Acceleration in the opposite direction destabilizes the flow. Knowing the direction of acceleration in a shear flow provides an estimate of whether rapid mixing will occur or whether the low-energy flow region will persist. Applied to impeller flow, for example, a realistic flow model becomes possible, as will be shown.

Vortex Flow

A vortex offers the most striking example of stabilized flow. In an equal-energy, potential vortex, the velocity varies in inverse proportion to the radius, $Vr =$ constant (Fig. 3.3). It will be observed that according to this relationship, when the radius approaches zero, the velocity approaches infinity. Therefore, such an ideal vortex cannot exist in real life. Indeed, real vortices possess a stagnant core near the center. This core rotates as a solid body with constant angular speed ω, and the velocity in the core becomes $V = \omega r$. The core boundary forms a shear layer. However, the centrifugal acceleration acts perpendicular to the streamlines, and points from the core, where lower energy prevails, radially out. Consequently, the shear layer is stabilized and can persist. More generally, it is found that if in a vortex with an arbitrary radial velocity distribution, anywhere the velocity decreases faster than $1/r$, an unstable shear layer exists. Such is the case, for example, at the wall of a cylindrical vessel containing a swirling fluid. The boundary layers are unstable and exceptionally fast mixing occurs. In the opposite case, when the velocity changes radially less fast than $1/r$, the flow is stable.

The core of a vortex is the only place where in real life stagnant fluid can remain stable and persist without mixing in a large flow field. In other cases, in wakes behind obstacles, the shear layers of the wake boundaries become unstable, generate irregular, unsteady flow fluctuations, and eventually, far downstream, mix with the main flow.

Separation

Further complications occur when the flow cannot follow the passage walls or flow boundaries and separates. Two typical cases of flow separation can be distinguished. The first type appears at sharp corners or sudden recesses in the wall, as shown in Fig. 3.4. The flow is unable to turn sharply around the corner, because the streamline curvature would become very small and the pressure rise perpendicular to the streamline very large. Since the fluid conserves its energy on the streamline approaching the corner, according to the Bernoulli equation, the velocity would have to become extremely high and the pressure at the corner would drop to theoretically infinitely low values. In real flow the streamline leaves

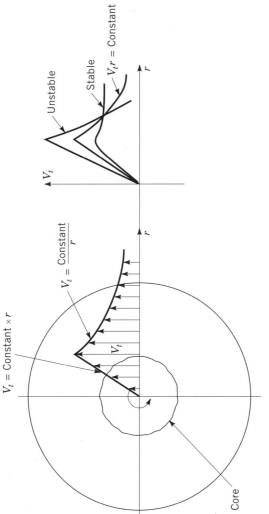

Figure 3.3 Potential constant-energy vortex: stable and unstable tangential velocity profiles.

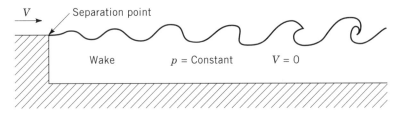

Figure 3.4 Backward-facing step; shear layer instability.

the wall at the corner and becomes a free streamline or shear layer. The region behind the corner becomes filled with stagnant or slowly moving, recirculating, low-energy fluid; it is a location where such low-energy fluid can accumulate and remain without being swept downstream. The shear layer is unstable and mixes farther downstream, resulting in an energy loss.

The second type of separation appears near gradually curving or even straight but diverging walls, where the passage cross section increases and the flow would have to slow down (Crighton 1985). With decreasing velocity, according to Bernoulli's equation, the pressure increases. Except for very special conditions, generally the flow cannot tolerate such a pressure increase and the streamline separates from the wall and becomes a free streamline, a shear layer. As before, low-energy fluid accumulates beyond the separation point near the wall. The shear layer mixes and some of the energy of the flow is lost. The downstream pressure cannot rise above its value at separation, which prevails throughout the separated region. Since the pressure along the free streamline, bordering the separated region, is constant, the flow velocity on the separated or free streamline must also remain constant. Such separation occurs, for example, behind obstacles placed into the fluid stream, or in diffusers, which are discussed below. Flow separation is responsible for major losses in pumping systems, and great effort is made to avoid them by streamlining the flow passages.

It is seen that the velocity and pressure distribution beyond the separation point depend on the location of the separation point, which unfortunately is sometimes difficult to determine. On gradually curving surfaces the velocity changes slowly ahead of the separation point, and secondary effects, such as those of viscosity, turbulence, and upstream conditions, can become important and can influence the exact location of the separation point. Refined computer calculation methods have been developed for determining separation points on airplane wings, for example. However, such calculations are not warranted for pump design, and empirical guidelines have been developed instead.

Two issues exist in pump design, where the onset and location of separation is particularly important and for which empirical guidelines have been found: flow in diffusers and flow through the impeller. These are discussed in subsequent chapters. It is important to remember that fluid flow can accelerate with very low losses, but always experiences losses when slowing down.

Lift

All objects placed in a fluid stream experience a pressure force, usually split into two components, one perpendicular to the main direction of flow, the *lift*, and a second component in the direction of the flow, the *drag* (Fig. 3.5). These forces are calculated by integrating the pressure on the surface of the object. Most calculation methods use inviscid potential or equal-energy flow assumptions to determine the pressure distribution on the surface of the object. However, the potential flow assumption does not allow for flow separation, which occurs on most objects, except on ideal airfoils at low incidence. Therefore, potential flow pressure distribution is assumed up to the location of the separation point, and constant wake pressure beyond. As a first approximation, separation can be assumed to occur where the velocity near the surface begins to decrease or slightly beyond.

Lift is particularly important for aircraft wings, but it appears on all objects, except on those having an axisymmetric (bullet- or droplike) shape, with their axis aligned in the direction of the flow. The lifting force can be attributed to rotation or circulation around the lifting body, which can be represented by a vortex with its axis perpendicular to the flow direction and also perpendicular to the direction of the lifting force vector. The rotational velocity due to the vortex adds to the main flow on one side of the body and opposes the main flow on the other. Therefore, the vortex induces higher velocities, and therefore lower pressures, on one side of the body than on the other, which results in a lateral resultant of the pressure force, a lift. Since according to fluid mechanic fundamentals a vortex filament cannot start or end in the free stream, the lifting vortex has to trail away from the wingtips or sides of the lifting body, leaving a wake in the form of two opposing vortices at equal distance, with their axes aligned in the direction of the flow, as shown in Fig. 3.6. Ideally, a vortex filament can only start at a stagnation point, where the velocity comes to zero. Such trailing vortices can often be observed on aircraft wingtips when landing in cloudy weather. Trailing vortices also exist behind road vehicles, for example (Sovran et al. 1978, Weihs 1980), and behind protrusions and holes on flow passage walls (Tuzson 1999). They are not visible in water, but are present in

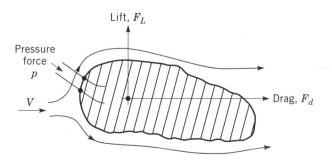

Figure 3.5 Lift and drag forces.

24   INSTABILITY, TURBULENCE, AND SEPARATION

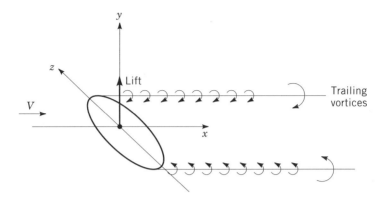

Figure 3.6   Three-dimensional lifting body and trailing vortices.

pumps, and result in serious erosion when the water carries sand, for example (Batchelor 1967).

Drag

Drag forces are responsible for most fluid energy losses. As described above, the flow separates from the surface of the body, and a low-energy wake appears behind the body. The wake is at approximately uniform pressure, equal to the pressure at the location of separation, and constant velocity persists on the wake boundaries. The shear layers at the wake boundaries become unstable. Irregular oscillations and eddies appear that mix the wake fluid with the main stream. The drag force F_d on various objects, having a cross section A, have been calculated theoretically and have been measured experimentally. Typically, the value of the drag coefficient C_d ranges from 0.1 to 1.0:

$$F_d = C_d A \rho \frac{V^2}{2}$$

Assuming a two-dimensional flow, if a control volume is defined around an object in a uniform stream by drawing boundaries at a great distance from the object, uniform free-stream conditions prevail on all boundaries except on the downstream boundary, where the wake trails away, as shown in Fig. 3.7. All forces acting on this volume must balance. Applying the force balance equation to this volume in the streamwise direction, the drag force on the object must be balanced by forces on the downstream boundary. If the low velocity of the fluid in the wake is parallel to the main stream, the pressure must be uniform and equal to the free-stream pressure on the downstream boundary. Therefore, the pressure forces cancel on all boundaries of the control volume. The momentum forces entering the volume, however, are greater than the momentum forces leaving,

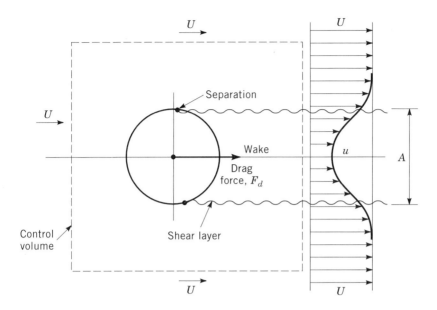

Figure 3.7 Drag force and wake behind an object in a uniform stream.

because of the velocity defect of the wake. This momentum defect must therefore balance the drag force of the object. The drag force of an object in a uniform stream can be measured by adding up or integrating the momentum defect across the wake (Batchelor 1967):

$$F_d = \rho U(U - u)A$$

Here u is the average velocity in the wake and A is the approximate cross-sectional area of the wake. If the wake contains trailing vortices, the pressure is not uniform on the downstream boundary of the control volume, and the corresponding pressure force must be added to the momentum force of the wake.

REFERENCES

Batchelor, G. K. (1967): *Fluid Dynamics*, Cambridge University Press, New York, pp. 348–353, 580–590.

Bayly, B. J., Orszag, S. A. (1988): Instability Mechanisms in Shear-Flow Transition, *Annual Review of Fluid Mechanics*, Vol. 20, pp. 359–391.

Crighton, D. G. (1985): The Kutta Condition in Unsteady Flow, *Annual Review of Fluid Mechanics*, Vol. 17, pp. 411–445.

Johnston, P. J., Halleen, R. M., Lezius, D. K. (1972): Effects of Spanwise Rotation on the Structure of Two-Dimensional Fully Developed Channel Flow, *ASME Journal of Fluid Mechanics*, Vol. 56, No. 3, pp. 533–557.

Sovran, G., Morel, T., Mason, W. T. (1978): Aerodynamic Drag of Bluff Bodies and Road Vehicles, *Symposium at the General Motors Research Laboratory*, Plenum Press, New York.

Tuzson, J. (1977): Stability of a Curved Free Streamline, *ASME Journal of Fluid Mechanics*, September, pp. 603–605.

Tuzson, J. (1993): Interpretation of Impeller Flow Calculations, *ASME Journal of Fluids Engineering*, September, pp. 463–467.

Tuzson, J. (1999): Slurry Erosion from Trailing Vortices, *ASME Fluids Engineering Conference*, San Francisco, FEDSM99-7794.

Weihs, D. (1980): Approximate Calculation of Vortex Trajectories of Slender Bodies at Incidence, *AIAA Journal*, Vol. 18, pp. 1402–1403.

4
PIPES, BENDS, AND DIFFUSERS

PIPES AND BENDS

For engineering calculations it can usually be assumed that the flow velocity in straight pipes is uniform across the cross section. A small boundary layer exists near the walls, for which allowance can be made by assuming a blockage of a few percent and a corresponding reduction in the cross-sectional area. The equation of continuity determines the velocity from the flow rate and cross-sectional area. If pipes branch or merge, the flow will split in a proportion of the cross-sectional areas of the branches.

In reality, upstream disturbances convect far downstream, and the establishment of uniform flow conditions requires about 10 pipe diameters of straight pipe run. Disturbances arise from piping components which change the direction or velocity of the flow: enlargements, bends, branches, tees, enlargements. These produce separation, trailing vortices, and turbulence, which extract energy from the flow and produce pressure losses. The lost energy is still present in the fluid in the form of irregular fluctuations and vortices, which only dissipate far downstream, but such irregular flow is of no use, and the corresponding energy represents a loss. Such losses in pipes and piping components have been measured. The loss is proportional to the velocity head of the flow. Loss coefficients are on the order of 1. More precise values for various piping components can be found in most engineering handbooks.

28 PIPES, BENDS, AND DIFFUSERS

The pressure loss in straight pipe runs is correlated in the form of a pressure coefficient C_p, or loss coefficient λ:

$$p_{02} - p_{01} = \lambda(L/D)\rho\frac{V^2}{2} = C_p\rho\frac{V^2}{2}$$

The curves shown on the Fig. 4.1 correspond to the correlation of Moody (1994) and Nicuradse (1933), which are also given by the following analytical expressions (Schlichting 1960, p. 525), where the Reynolds number corresponds to $\text{Re} = DV/\nu$:

$$\lambda = \frac{64}{\text{Re}} \quad \text{for laminar flow}$$

$$\frac{1}{[2\log(0.5D/k) + 1.74]^2} \quad \text{for turbulent flow}$$

In this expression k is the average size of the surface roughness elements. As an example, typical cast iron pipes have a roughness height of $k = 0.01$ in $= 0.25$ mm. Turbulent flow usually exists in those piping components that operate at Reynolds numbers Re beyond 100,000 (note that "log" stands for the base 10 logarithm) (Hydraulic Institute 1961).

In the expression above for the pipe friction loss, L is the length of the pipe and D the diameter. If the pipe cross section differs from the circular shape, D becomes the hydraulic diameter D_h, which corresponds to four times the cross-sectional area A divided by the circumference C: $D_h = 4A/C$. Losses of piping components are specified by an appropriate pressure coefficient C_p. Practical calculation procedures express the pressure drop from piping components as the pressure drop from an equivalent length of straight pipe, as shown in Table 4.1. Typical flow velocities in pipes range from 5 to 25 ft/sec (1.5 to 7.5 m/s).

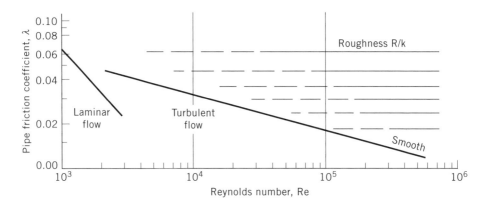

Figure 4.1 Pipe friction coefficient as a function of Reynolds number.

Table 4.1 Equivalent pipe length of piping components for estimating losses

Piping Component for Pipe Diameter D	Equivalent Length of Pipe (in.)					
	1	2	4	8	16	32
Elbow 90°						
Regular	1.6	3.1	5.9	12	21	42
Long radius	1.6	2.7	4.2	7	10	17
Elbow 45°	0.8	1.7	3.5	7.7	15	
Return bend 180°	1.6	3.1	5.9	12	21	
Valve						

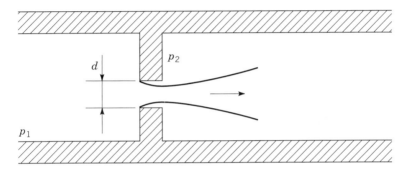

Figure 4.2 Orifice flow and pressure drop.

In case of constrictions or orifices, the pressure drop is often known, and the flow rate Q is to be calculated, d being the orifice diameter (Fig. 4.2).

$$Q = C_d \frac{\pi}{4} d^2 \left[\frac{2(p_1 - p_2)}{\rho} \right]^{1/2}$$

Here C_d is a discharge coefficient that accounts for constriction of the streamlines in the orifice and for the variation of the velocity across the cross section. For sharp-edged orifices the discharge coefficient is on the order of 0.7. With careful rounding it can approach 1.

Reaction forces on 90° bends can be calculated by applying the momentum equation in the direction of the incoming and exiting flows separately, since the forces in the two directions are independent (Fig. 4.3). The reaction forces must be resisted either by the piping connections or by external supports. The direction of the resulting external force, the vectorial sum of the forces in the two directions, is at 45° to the direction of the two components.

In either direction the reaction force is

$$F_x = (\rho V^2 + p) A$$

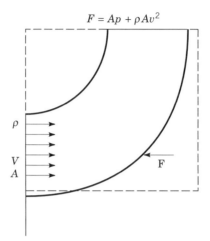

Figure 4.3 Reaction force on a pipe bend.

In this expression V is the flow velocity, p the local static pressure at the inlet or exit flange, and A the cross-sectional area of the pipe. Water hammer pressure pulses can generate considerably larger, instantaneous reaction forces and should be considered. Various standards regulate the required strength of piping connections in practical applications.

DIFFUSERS

A great portion of the energy imparted to the fluid by pumps consists of a velocity head: $\rho V^2/2$, whereas the energy is usually desired in the form of pressure. Consequently, an important component of pumps is a diffuser, which transforms the velocity head leaving the pump impeller into a pressure head by slowing down the flow gradually. As explained in Chapter 3, the flow has a strong tendency to separate when it encounters a pressure rise. It takes a very careful design to avoid separation and achieve a significant pressure recovery in a diffuser.

Diffusers achieve pressure recovery or diffusion by gradually increasing the flow passage cross section (Fig. 4.4). According to an empirical rule of thumb, in straight-wall diffusers the walls should diverge by an angle of about 7°. More precise, but still empirical guidelines have also been developed (Carlson et al. 1967, Johnston 1998, Muggli et al. 1997, Reneau et al. 1967) and are shown in Fig. 4.5. Experiments show that the least losses occur when the flow is on the verge of separating. In practice, inlet disturbances and a nonuniform inlet velocity distribution require a more conservative design.

DIFFUSERS   31

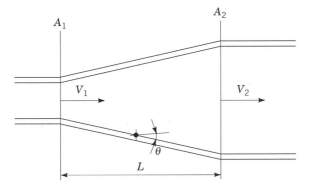

Figure 4.4 Diffuser.

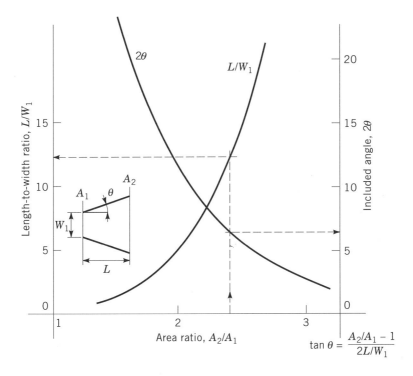

Figure 4.5 Preferred diffuser geometry.

Diffuser performance is illustrated by plotting the pressure recovery C_p, which is the fraction of inlet velocity head converted to pressure head, against the length of the diffuser or against the increasing cross-sectional area. The performance of actual diffusers can be bracketed on this plot by two theoretical curves: the perfect loss-free diffusion and the sudden expansion loss, which can be considered (with some qualification) the worst case.

32   PIPES, BENDS, AND DIFFUSERS

The ideal, loss-free case can be calculated from the continuity and energy equations applied to a straight two-dimensional diffuser.

Continuity:

$$V_1 A_1 = V_2 A_2$$

Energy:

$$\rho \frac{V_1^2}{2} + p_1 = \rho \frac{V_2^2}{2} + p_2$$

Substituting $V_2 = V_1(A_1/A_2)$ and rearranging yields

$$C_p = \frac{p_2 - p_1}{\rho V_1^2/2} = 1 - \left(\frac{A_1}{A_2}\right)^2$$

The sudden expansion case corresponds to the flow configuration when a flow passage suddenly opens into a much larger one, as shown in Fig. 4.6. The flow separates and a jet issues into the larger passage. As the direction of velocity is along the straight flow passage, in this case the continuity equation and the force balance equation in the direction of the velocity can be used to obtain a solution, provided that the flow is uniform and completely mixed at the end of the larger passage, at the cross section A_2. The passage configuration and the boundaries of the control volume are shown in Fig. 4.6. Note that the pressure in the separated region equals the pressure at the separation point of the flow.

Continuity:

$$V_1 A_1 = V_2 A_2 \quad \text{or} \quad V_2 = V_1 \frac{A_1}{A_2}$$

Force balance:

$$\rho V_1^2 A_1 + p_1 A_1 + p_1(A_2 - A_1) = \rho V_1^2 A_2 + p_2 A_2$$

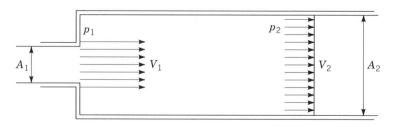

Figure 4.6 Sudden expansion.

Substituting V_2, dividing by A_2, and rearranging yields

$$C_p = \frac{p_1 - p_1}{\rho V_1^2/2} = 2\left[\left(\frac{A_1}{A_2}\right) - \left(\frac{A_1}{A_2}\right)^2\right]$$

The pressure recovery of practical diffusers typically falls between these two cases (Fig. 4.7). It can be less than the pressure recovery for sudden expansion if the diffuser is not long enough and the flow is not completely mixed at the exit. Practical diffusers rarely exceed area ratios greater than 2. At an area ratio of 2, ideally 75% of the inlet velocity head could be recovered, while a sudden expansion would still recover 50%. Most smaller pumps use short diffusers to save space, and only 50% or less of the velocity head is recovered. Also, the flow velocity at the diffuser inlet is nonuniform. It leads to early separation, beyond which no further kinetic energy can be recovered.

Since both the pressure and velocity at the diffuser inlet and exit are known, the equation of conservation of energy, Bernoulli's equation, can be used to calculate the energy lost between the inlet and exit. Experiments with curved diffusers lead to the conclusion that turning through 30° or less leaves the pressure recovery essentially unaffected (Fox and Kline 1962).

In diffuser design the greatest attention should be given to the front portion of the diffuser, where the velocity is still high, and where disproportionately more pressure can be recovered. For example, in a straight-walled diffuser about half of the energy is recovered in the first third of the diffuser. In some centrifugal pumps a radial diffuser section immediately follows the exit of the impeller. The radial diffuser can consist of passages formed by airfoil-shaped guide vanes arranged in a circle, or of several straight, diverging diffusers. In Chapter 11 we offer details on the design of diffusers in pumps. Diffuser loss accounts for the greatest part of

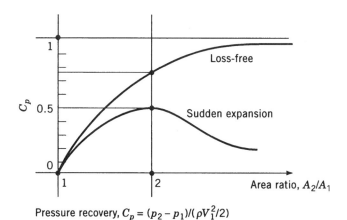

Figure 4.7 Pressure recovery coefficient versus area ratio for a loss-free diffuser and for a sudden expansion.

pump losses, and therefore its design offers the greatest challenge in pump design.

Obstructions in Pipes

The sudden expansion flow model can also be used to estimate the pressure drop from objects obstructing the flow in pipes. Only the net flow cross sections enter into calculation of the sudden expansion loss; their particular configuration remains unspecified. Therefore, the net free cross-sectional area around the object can be taken as the inlet cross section A_1. In the case of sharp-cornered objects, such as a flat plate perpendicular to the flow, allowance must be made for a contraction of the flow moving past the object and the cross-sectional area must be multiplied by a contraction coefficient. Depending on the rounding of the corners of the object, the contraction coefficient may vary from 0.7 to 1.0. As before, a condition for the validity of the sudden expansion flow model is that the flow be fully mixed and the velocity be uniform at the downstream cross section.

Such an obstruction loss model is sometimes applied to the specific case of the blade wake mixing loss behind blade rows. Blades and vanes have some definite thickness and leave a wake in the flow behind them. It can be assumed that the cross section of the flow passage between the blades of the blade row suddenly increases by the amount of the blade thickness as the flow leaves the blade row. Since the blade thickness is considerably less than the passage width, the area increase and the corresponding loss are relatively small and may or may not be negligible. A sudden expansion loss calculation can provide an estimate of the magnitude of such a wake mixing loss.

Although sometimes applied to rotating impellers, such a wake mixing loss estimate should only be applied to stationary blade rows. The flow leaving the impeller is unsteady when considered by a stationary observer in an absolute frame of reference. The absolute flow velocity is not parallel but approximately perpendicular to the wake. The mixing process is therefore entirely different from a steady wake mixing. The fluid in the wake is not low-energy fluid as in the case of a wake behind a stationary object, but can move with a higher velocity (approximately with the tip speed) than the main flow. In addition, trailing vortices are likely to be present in the corners between the blade and the hub and shroud surfaces. Very fast mixing has been observed in tests, resulting in surprisingly low losses. The explanation was offered that energy transfer takes place by normal, unsteady pressure forces—by pressure exchange (Foa 1958) rather than by viscous or turbulent mixing.

REFERENCES

Carlson, J. J., Johnston, J. P., Sagi, C. J. (1967): Effects of Wall Shape on Flow Regimes and Performance in Straight, Two-Dimensional Diffusers, *ASME Journal of Basic Engineering*, March, pp. 151–160.

Foa, V. J. (1958): Pressure Exchange, *Applied Mechanics Review*, December, pp. 655–657.

Fox, R. W., Kline, S. J. (1962): Flow Regimes in Curved Subsonic Diffusers, *ASME Journal of Basic Engineering*, September, pp. 303–312.

Hydraulic Institute (1961): *Pipe Friction Manual*, HI, New York.

Johnston, J. P. (1998): Review: Diffuser Design and Performance Analysis by a Unified Integral Method, *ASME Journal of Fluids Engineering*, March, pp. 6–12.

Moody, L. F.(1944): *Friction Factors for Pipe Flow*, Vol. 66, AMSE, New York, p. 671.

Muggli, F. A., Eisele, K., Casey, M. V., Gulich, J., Schachenmann, A. (1997): Flow Analysis in a Pump Diffuser, Part 2: Validation and Limitations of CFD for Diffuser Flows, *ASME Journal of Fluids Engineering*, December, pp. 978–985.

Nicuradse, J. (1933): Strömungsgesetze in rauhen Rohren, *VDI Forschungsheft* No. 361.

Reneau, L. R., Johnston, J. P., Kline, S. J. (1967): Performance and Design of Straight, Two-Dimensional Diffusers, *ASME Journal of Basic Engineering*, March, p. 141.

Schlichting, H. (1960): *Boundary Layer Theory*, McGraw-Hill, New York.

5

PUMPING SYSTEMS

PUMP SPECIFICATIONS AND SELECTION

Before pump design or selection can begin, specifications need to be established which express several requirements. Only requirements pertaining to the hydraulic performance of the pump are considered here. Other considerations are reliability, durability, noise and vibration, maintenance, installation, and control. These concerns affect primarily the mechanical design of the pump: material selection; auxiliary component specifications, such as bearings, seals, and instrumentation; and design for ease of assembly, service and maintenance.

Among the specifications affecting hydraulic performance, the most basic are the nominal efficiency, speed, flow rate, pressure or head, and power. Efficiency is important from the point of view of energy cost. Because of the maturity of pump design technology, the nominal efficiency of commercially available, competitive pumps is within 1 or 2% test data rarely being more accurate than 1%. Power losses arising from piping system design and losses due to control considerations can amount to 30% or more of the energy supplied by the pump. In process applications pumps are often followed immediately by control valves, which produce a head loss of 20 or 30% at nominal conditions. Flow control is achieved by increasing or decreasing the head loss. Considerable energy savings can be achieved by variable-speed drives, for example. Consequently, energy conservation efforts should preferably be directed at these losses rather than at 1 or 2% pump efficiency improvements. The competitive advantage of high efficiency therefore appears to reside in reflecting high technical competency on the part of the manufacturer rather than in energy savings.

If the power of the drive motor is limited, nominal efficiency might also be a consideration. Depending on the application, motors are usually selected assuming a certain service factor, which multiplies the nominal power requirement of the driven equipment. A factor greater than 1 assures a reserve power margin. Actually, maximum temperature rise limits the ultimate power that an electric motor can deliver. In instances where the pump and motor are sold as an assembly, as in appliances for example, pump and motor are closely matched, assuming a service factor of 1. The motor load can reach the ultimate power, limited only by the permissible temperature rise. In such a case the pump efficiency determines the maximum pressure or flow rate that can be obtained from the assembly with a given motor.

Pump speed will depend primarily on the driver. Electric motor speeds follow the network frequency: 60 Hz in the United States and 50 Hz in Europe, corresponding to nominal 3600- and 3000 rpm synchronous speeds of single-pole motors. Smaller motors, up to perhaps 100 hp, usually turn at 3600 rpm. Larger motors turn at an even fraction of the synchronous speed: 1800, 1200, 900, or 720 rpm, depending on the number of poles of the motor.

Large pumps have large inlet diameters. The impeller inlet circumferential velocity increases with the diameter but must stay in reasonable proportion to the inlet flow velocity to avoid blade angles too much inclined toward the tangential direction. The danger of cavitation also increases with the velocities at the inlet. These reasons suggest lower rotational velocities for large pumps. Under full load, motor speed typically decreases 2% from the synchronous speed. At a price, variable-frequency drives are available, some of which can reach 5400 rpm, but the synchronous speed limits the maximum speed of most induction motors. Internal combustion engines drive pumps in remote, outdoor, often mobile applications, such as irrigation, dewatering, and flood control. Their rated speed varies but remains relatively low compared with that of electric motors, on the order from 1000 to 2000 rpm. Their speed is usually governed to remain constant regardless of the load.

The rated flow rate of pumps can match the required flow directly since leakage losses are normally insignificant. The nominal flow rate also determines the inlet and outlet pipe diameter and flange size, which are standardized. The code number of commercial lines of pumps often includes the inlet and outlet pipe diameters measured in inches. Pipe and flange diameters are sized to result in water flow velocities of 5 to 15 ft/sec (1.5 to 5 m/s), but velocities can reach 30 ft/sec (10 m/s), in large high-specific-speed pumps.

Specifying the pressure or head rise produced by the pump requires the most care. It is defined as the total head, static and dynamic, at the pump exit flange minus the total head at the inlet flange. The nominal head required by the system must be estimated by keeping in mind the maximum possible head that might ever be encountered. A reserve margin may be necessary. However, an excessive margin may make the pump operate normally at low efficiency or at some adverse condition. Most pumps are capable of operating over a range of higher or lower flow rates and pressures on either side of the nominal. Such off-design operation, at exceptional conditions, should be taken into account when estimating the nominal head, in order to minimize the need for a large reserve margin.

The head rise required from the pump is estimated by adding up the suction head at the inlet, including pipe friction losses, valve and pipe friction losses at the exit, and the net head of the final load, such as the elevation head to an overhead tank (Fig. 5.1). If the pump inlet is below the water level of a sump or supply tank, the inlet head will be positive, from which the inlet piping losses, calculated at the rated flow rate, need to be subtracted. If the pump inlet is above the water supply level, the inlet head is negative and so are the additional inlet piping losses. The valve and pipe losses at the exit, also calculated at the rated flow rate, as well as the net head of the load, add to the head requirement of the pump.

Particular attention must be given to the pump exit velocity head. The rated pump head rise usually includes the velocity head of the flow leaving the pump. In performance testing the pressure rise is measured "total to total," implying that the velocity heads at the inlet and exit flanges are added to the respective static head measurements to arrive at the total head at each location. This approach assumes that the kinetic energy of the flow leaving the pump remains useful, which may or may not be the case. If the pump produces a free jet, like a fire

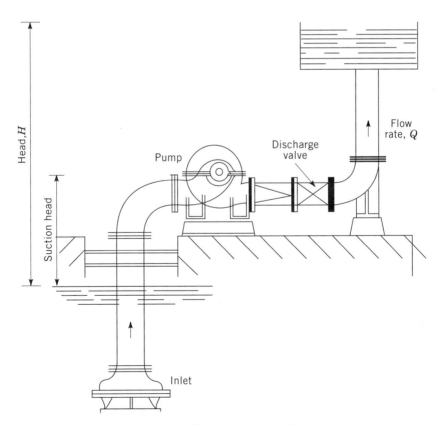

Figure 5.1 Pump system installation.

hose, then the velocity head can indeed be counted as useful energy. If the pump discharges into a stagnant tank, for example, the kinetic energy is lost and must be added to the piping losses at the pump exit.

In addition to specifying the rated efficiency, speed, flow rate, and head, several other requirements need to be established. First, the pump must have sufficient suction capability without cavitating. Verifying satisfactory suction performance requires two steps: the suction head—the net positive suction head available (NPSHA) from the system—must be estimated, and a comparison must be made with the net positive suction head required (NPSHR) by the pump. Pump catalogs list the NPSHR of the pump as a function of flow rate. The NPSHR values reach a minimum at the design flow rate and gradually increase at higher flow rates. The absence of cavitation must be assured at the greatest anticipated flow above the rated flow rate. Catalog data are usually obtained at conditions conforming to testing standards, typically when the pump head declines by 3% with decreasing inlet pressure. However, cavitation vapor bubbles may appear earlier and cause damage to the impeller. Conditions in the actual application may differ from the standard testing conditions. Excessive aeration of the water ahead of the pump, supply sump vortexing, or air leaks into the suction piping may cause pump cavitation at inlet pressures that would appear sufficiently high according to catalog data. Therefore, the NPSH required should be estimated conservatively, and a higher inlet pressure than that absolutely necessary should be provided to the pump.

Often, further requirements are imposed, such as the minimum flow rate at which the pump may have to function, the head at shutoff, and the maximum power demand. Several undesirable phenomena can appear at reduced flow rate: general instability of operation, inlet recirculation, exit recirculation, cavitation, and overheating. Details of such conditions are discussed in subsequent chapters. Some of these can be tolerated in case of an emergency but would lead to deterioration and failure of the pump if they were to persist for longer periods. The application will decide how often and how long the pump may operate at reduced flow and what the cost trade-off might be between a more expensive pump and damage from an emergency of low probability. Manufacturers offer guidelines for the safe, minimum flow of their pumps (Cooper 1988).

Instability at reduced flow results from a positive slope of the head–flow curve of the pump. It depends not only on the pump characteristics but also on the system load curve, as described further below. It causes severe torque and power fluctuations as well as vibration and noise, which can damage the bearings or shorten their life. Inlet and exit recirculation may or may not create problems. It certainly disrupts the normal flow pattern in the pump, but can often be tolerated for a limited time. The flow rate corresponding to the onset of inlet and exit recirculation can be calculated with the procedures and computer programs given in this book. Recirculation can also combine with cavitation, which might aggravate vibration and power fluctuations, and will result in progressive cavitation damage. Recirculation is more likely to precipitate cavitation if the calculated NPSHA does not comprise a sufficient safety margin.

Overheating results when excessive losses, transformed to heat, are retained in the pump. Pumps handling cold water rarely encounter problems. However, a temperature rise will affect the vapor pressure of the fluid. When pumping hot fluids near boiling, cavitation may appear unexpectedly. The head at shutoff is difficult to estimate because the flow churns completely uncontrolled in the pump casing. It also depends on the configuration of the piping upstream from the pump. An approximate estimate can be obtained by extrapolating the performance curve to zero flow. The pressure at shutoff is specified, for example, in the case of fire pumps. Such a specification may intend to make certain that the head–flow curve does not have a range of positive slope, which would lead to instability.

If the power demand of the pump decreases with increasing flow rate, a reserve power margin will be needed if the pump is likely to operate below the flow rate at the rated point. The opposite holds for pumps whose power demand generally increases with the flow rate. Pump characteristic plots in catalogs show the power demand of the pump as a function of the flow rate. If the plot does not show the power directly, it can be calculated from the head, flow rate, and efficiency.

SYSTEM MODELING

The desired performance of a pump in a pumping system defines its design. The need for a pump arises from an overall system design, the purpose of which is to convey some fluid from one location to another. Proper functioning of the pump and the overall system must be ascertained before the hardware is procured and the system is built. Thought must be given not only to steady-state performance, but also to time-dependent variation and sudden changes in speed, torque, flow rate, and pressure, as for example during startup, shutdown, or sudden valve closure. The pumping circuit might be part of a process, which will need to be controlled automatically. Since the pump feeds energy into the system, it can also power undesirable instabilities, which must be avoided. A mathematical modeling of the pumping system may be desirable to test its proper functioning.

System modeling and analysis can also be used as a diagnostic tool. Electric power variations or disturbances can reveal mechanical defects in deep-well pumps, which are not accessible to direct scrutiny (Kenull et al. 1997). High-speed computers and commercially available computer programs have considerably facilitated the steady and unsteady mathematical analysis of pumping systems. However, the modeling of particular systems still demands skill. Excessively simplified models provide only trivial answers. Excessively complex models become unwieldy, requiring much labor and cost.

Pumping system components transform and transmit energy which cannot be lost and must be accounted for. Typically, electric energy, taken from the grid, is transformed to shaft work by an electric motor; the pump follows, which transforms the shaft work into hydraulic energy; the valving and piping system transmits the hydraulic energy while also producing some losses; finally, the

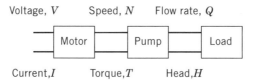

Figure 5.2 Pump system model.

energy is stored or absorbed by the load (Fig. 5.2). Since it is the same energy that appears in different forms in these components, the variables, which characterize them, show certain similarities and suggest a unified mathematical approach (Blackburn et al. 1960).

The discipline of control theory has developed practical modeling methods to characterize those systems that were to be controlled (Blackburn et al. 1960). A similarity exists between electrical, mechanical, hydraulic, and even thermal systems. Before the widespread use of high-speed digital computers, system analysis was often performed on analog computers, where electric models of mechanical, hydraulic, or thermal systems were assembled using the electric equivalents of such components. A numerical analysis is now possible with modern digital computers.

Each component can be modeled by a black box with two input ports, corresponding to two input variables, and two output ports, corresponding to the two output variables. Such four-terminal element models originated in the analysis of electronic systems. Here, applied by analogy to pumping systems, the dimension of the input and output variable pair must be such that their product becomes a variable having the dimension of power. For example, in the case of the electric motor, voltage and current are the input variables and shaft speed and torque are the output. In pumps, shaft speed and torque are the input and flow rate and pressure are the output. In the case of a hydraulic brake shaft, speed and torque are the input, and mass flow rate and the temperature of the cooling water, or the heat capacity per unit mass of fluid, are the output. Since no perfect components exist, most have a sidestream of lost energy, usually in the form of heat. The output power remains less than the input power, their ratio being the efficiency of the component.

The input variables, x_1 and y_1, and output variables, x_2 and y_2, are related, by two equations, for example by f_2 and g_2:

$$x_2 = f_2(x_1, y_1)$$
$$y_2 = g_2(x_1, y_1)$$

Since there are four variables and two equations, two variables must be given to determine the two others. The two equations are derived from governing physical principles. One of the equations might state that the input power must be equal to the output power and the losses. This equality expresses the conservation of energy. Other equations are usually obtained from force balance equations in the case of mechanical systems.

If the equations are linear, the output variables can be expressed explicitly in terms of the input variables alone:

$$x_2 = a_{11}x_1 + a_{12}y_1$$
$$y_2 = a_{21}x_1 + a_{22}y_1$$

In such a linear form the equations are easily manipulated. Any two of the variables can be calculated in terms of any two others. The four coefficients, some of which might be zero, fully characterize the particular system element. For example, many electrical components, consisting of inductances, resistances, and capacitances, and many mechanical components, consisting of masses, viscous friction elements, and springs, yield linear equations.

The torque–speed characteristics of electric motors, which correspond to one of the needed equations, can differ greatly (Fig. 5.3). In general, the torque of commercial induction motors remains approximately equal to the nominal torque up to a speed of about 70% of the nominal speed. It rises rapidly to about 200% of nominal torque at 80% speed and drops to zero at the speed, which corresponds exactly to the network frequency. If the motor starts a loaded pump, instability problems might arise around 70% speed. Large pumps need special

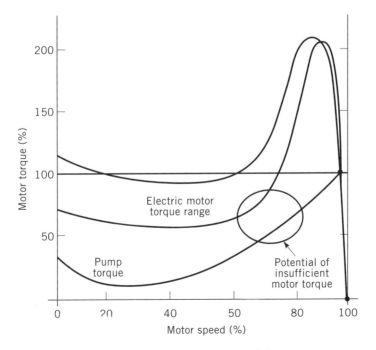

Figure 5.3 Electric motor torque–speed diagram.

starting procedures (Sulzer 1989). Once the motor has come up to speed, the motor acts very stiff; large torque or power variations result in little speed change. Often, the speed can be assumed constant and independent of the torque for modeling purposes.

In the case of pumps, the input variables are the shaft speed N and torque T, and the output variables are the head H and the flow rate Q. The first equation corresponds to the head–flow curve for a given shaft speed. The flow rate and speed being given, the head–flow curve determines the output head. The other output variable, the flow rate, can be calculated from the input power multiplied by the efficiency and divided by the head. Theoretically, if any two of the variables are given, the other two can be calculated. As will be seen, the two given variables will be defined by the components preceding and following the component under consideration.

$$H = H(Q, N, \eta)$$
$$Q = Q(H, T, N, \eta)$$

In terms of the actual variables—speed N (rpm), torque T (ft-lb/sec), head H (ft), flow rate Q (ft^3/sec), efficiency η, constant geometrical quantities diameter D_2, exit width B_2, slip coefficient of the pump σ, exit blade angle β_2, and gravitational acceleration $g = 32.2$ ft/sec^2 (9.8 m/s^2)—the two equations defining the steady-state output of a pump become

$$H = \eta \left(\frac{\pi D}{60}\right)^2 \frac{\sigma}{g} N^2 - \eta \left(\frac{\pi D}{60}\right) \frac{\tan \beta_2}{\pi D B g} QN$$

$$Q = \frac{\eta}{\rho g} \frac{NT}{H}$$

The head–flow equation above corresponds to the form used in the performance calculation method of this book, which is given in Chapter 10, where all variables are defined.

$$H_{\text{th}} = \frac{H}{\eta} = \frac{U(U\sigma - W_r \tan \beta_2)}{g}$$

Ultimately, this equation derives from application of the principle of conservation of angular momentum to pump impellers. The following substitutions lead to the equation above: $U = \omega D_2/2$, $\omega = 2\pi N/60$, and $W_r = Q/\pi D_2 B$.

The second equation corresponds to the principle of conservation of energy:

$$\text{power} = TN = \frac{QH\rho g}{\eta}$$

These equations are nonlinear and contain implicitly the head H and flow rate Q, as well as the efficiency η, which depends on the flow rate. Unlike many electrical components, the practical relationships, characterizing hydraulic components, are mostly nonlinear, and therefore the variables cannot be calculated explicitly. They would have to be solved by trial and error, by progressive approximations. However, powerful computer programs are available today that can directly solve even nonlinear, implicit equation systems.

A further complication arises when the equations are time dependent: for example, when the time derivatives or integrals of the variables appear in the characteristic equations. Such cases occur when inertia or compressibility forces appear (Blackburn et al. 1960, Vance 1988).

To illustrate calculations, the case of a rotating coupling with inertia I and a torsional windup spring rate k is considered here (Vance 1988) (Fig. 5.4). The input T_1 and output T_2 torques are equal except for an inertia torque, which is proportional to the time rate of change of the input speed dN_1/dt. The difference between the input speed N_1 and the output speed N_2 is equal to the shaft windup, which is proportional to the time rate of change of the input torque dT_1/dt:

$$T_2 = T_1 - I\frac{dN_1}{dt} \qquad N_2 = -\frac{1}{k}\frac{dT_1}{dt} + N_1$$

Since the relationships happen to be linear, one can identify the coefficients a using the symbolic derivation d/dt as

$$a_{11} = 1 \qquad a_{12} = -I\frac{d}{dt} \qquad a_{21} = \frac{1}{k}\frac{d}{dt} \qquad a_{22} = 1$$

The inertia of practical couplings used between skid-mounted electric motors and pumps can become significant, depending on the rate of change of the shaft speed. Catalogs of commercial couplings often list their weight and inertia. Couplings are made somewhat compliant to absorb torsional shaft vibrations and misalignment. However, their total torsional windup, from no load to rated torque capacity, which can be used to calculate the spring rate, amounts to only a few degrees, and the couplings can usually be considered rigid.

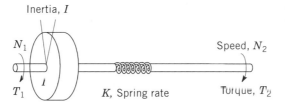

Figure 5.4 Model of shaft with inertia and torsional spring.

46   PUMPING SYSTEMS

In the case of pumps the inertia of the impeller must be taken into account and must be added to the torque load when shaft speed changes are expected. The mass of the fluid contained in the impeller is combined with the mass of the metal to calculate the rotational inertia. The polar inertia I of a disk of diameter D and thickness L or weight W has the dimension of lb/ft/sec^2 in English units:

$$I = \rho\left(\frac{\pi}{32}\right)D^4 L = \frac{WD^2}{8g}$$

The impeller is fixed rigidly to the shaft, and the compressibility of the liquid contained in the pump is negligible. Therefore, no shaft windup or compliance needs to be introduced. The pump delivers to a load, which will typically consist of an exit pipe, a valve or orifice and a tank at a higher elevation. A long pipe will represent a steady friction load and an inertia load of the fluid contained in it. A valve or orifice, generates an energy loss, and an elevated tank will store energy.

Transient, unsteady operation of the pump, such as startup, shutdown, and loss of power, can drive the pump into unusual operating ranges, such as reversed flow or reversed rotation. A full, time-dependent analysis of such upset conditions often requires that the pump characteristics be known in these unusual ranges of operation. Since the speed may change continually, extensive use is made of the similarity relationships by normalizing the flow rate, head, and torque with the rotational speed. Frequent use is also made of a four-quadrant representation of the performance in the four fields of positive and negative flow and speed. Other representations have been proposed which are better suited to the computer calculation of pump transients (Martin 1983).

The need to model and calculate the response and transient, dynamic behavior of long pipelines has generated specialized analytical methods (Wylie and Steeter 1993). Detailed discussion of the analysis methods would exceed the scope of this book and cannot be covered here. Pressure wave propagation and water hammer calculations have become prerequisites of pipeline and power plant construction. The complexity of most piping systems usually requires that a computer be used for its analysis. If relatively short piping connections are used, simplified models suffice for a first, approximate check. Steady pipe friction losses vary with the square of the flow rate. The inertia load corresponds to the mass of the fluid in the pipe and varies with the time rate of change of the flow rate. The compressibility of the fluid can usually be neglected. When the pipe is very long, the pipe walls stretch easily, or when air pockets exist somewhere, the volume capacity of the piping system increases with increasing pressure, which has the same effect as compressibility of the fluid. Under such conditions, which are difficult to quantify, resonant pressure oscillations can be excited by the pump. The most frequently encountered frequency of excitation is the blade passing frequency of the pump—the number of shaft revolutions per second times the number of impeller blades.

The piping and load downstream of the pump are characterized by expressions of the form

$$H_2 = H_1 - h - kQ_1^2 - \frac{L}{gA}\frac{dQ_1}{dt}$$

$$Q_2 = Q_1$$

In the expression of the output head, H_2, which is the head remaining at the pipe exit, the static elevation of the tank h, the pipe friction and valve or orifice loss, and the inertia is deducted from the input head, H_1, which in this instance designates the head at the pump exit. This equation corresponds to the conservation of energy. The second equation assumes that there are no leaks and that the input flow is equal to the output flow, which corresponds to the continuity equation. If the pump delivers into a tank open to ambient pressure, the output head H_2 is a known constant and the expression relates the input head H_1 and flow rate Q_1, which correspond to the pump output head and flow rate.

For the sake of illustration we consider steady operation of a pump, which is driven at constant speed N and delivers the flow Q into an elevated tank at constant head h throttled by a valve (Fig. 5.5). To calculate the four variables defining the operation of the pump—N, T, H, and Q—four equations are now available. The shaft speed N is known to be constant. The head H and flow rate Q can be determined from the head–flow characteristics of the pump and from the characteristics of the load. Finally, the input torque T is obtained from the expression equating the input and output power and losses.

A graphic solution can best illustrate the calculation of the head and flow rate of the operating point. The characteristic curve of the pump on a head versus flow rate plot is typically convex in the direction of increasing head, as shown in Fig. 5.5. The slope of the head–flow curve is usually negative. The head decreases with increasing flow rate. The load curve corresponds to a parabola with its origin

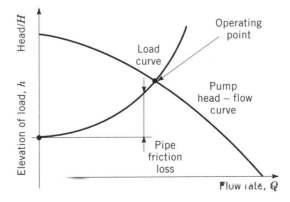

Figure 5.5 Head versus flow rate plot showing pump and load characteristics.

on the vertical axis at a value of h. The slope of the load curve is usually positive. The intersection of the two curves gives the operating point H and Q of the pump.

Let us now assume that the head–flow curve of the pump has a maximum. The head decreases in both directions, with increasing and decreasing flow rate, from its value at the maximum head. Let us further assume that the pump gradually fills a tank in which the water-level elevation h rises slowly: $h = H$, and that the head loss in the piping is negligible. The load curve, which now consists of a horizontal line, intersects the head–flow curve at two points on either side of the maximum, as shown in Fig. 5.6. Both are possible operating points, but as it turns out, only the one at high flow rate remains stable.

The stability of an operating point can be determined by slightly disturbing the location of it and observing whether the pump tends to come back to its steady operating point. If the operating point at the lower flow rate were to be displaced to slightly larger flow rate, the head produced by the pump would also rise slightly. The excess head would overcome inertia and accelerate the flow rate to even larger values. The operating point would move away and would not come back to the original operating point. Therefore, this operating point will be unstable. On the other hand, if the operating point at the larger flow rate were moved slightly higher, the pressure produced by the pump would fall and the flow would tend to slow down and come back to the original operating point. Therefore, the operating point at the lower flow rate is unstable, and the operating point at the higher flow rate is stable. It will be observed that a stable operating point exists if the slope of the load curve is greater than the slope of the head–flow curve.

Such a stability analysis can also be performed on a computer. Dynamic system models are disturbed by imposing a small perturbation, usually a small sinusoidal oscillation, added to one or the other of the variables, and by calculating whether the amplitude of the oscillation would grow or die out. Although this analysis method may appear artificial, it can be reasoned that some

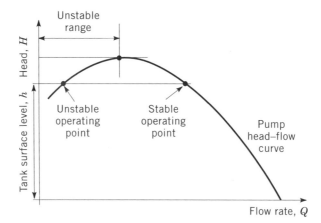

Figure 5.6 Head versus flow rate plot showing pump characteristics with unstable range.

unwanted disturbances will always be present in real life. A system in an unstable equilibrium position will leave it and never return. Theoretically, an egg can be stood on end, but in practice it will never remain standing. A system will eventually return to a stable equilibrium position regardless of the manner in which it is disturbed, and regardless of the path it takes to return.

As the elevated tank continues being filled, eventually the head will reach the value corresponding to the maximum of the head–flow curve. The pump cannot produce higher pressure and the flow will stop. The equivalent operating point will jump to the vertical axis. It may even surge back and forth between no flow and the maximum operating point if the tank slowly drains. The pump will never be able to operate at flow rates below that which corresponds to the maximum head. Because of such unstable surging, pumps having a head–flow curve with a positive slope and a maximum should be avoided for general use. In Chapter 14 we discuss the effect of impeller exit recirculation on the presence of a maximum on the head–flow curve and on the stability of the pump at a reduced flow rate.

REFERENCES

Blackburn, J. F., Reethof, G., Shearer, J. L. (1960): *Fluid Power Control*, MIT Press, Cambridge, Mass., pp. 130–143, 433–497.

Cooper, P. (1988): Panel Session on Specifying Minimum Flow, *Proceedings of the Texas A&M 5th International Pump User Symposium*, Houston, Texas, pp. 177–192.

Imaichi, K., Tsujimoto, Y., Yoshida, Y. (1982): An Analysis of Unsteady Torque on a Two-Dimensional Radial Impeller, *ASME Journal of Fluid Mechanics*, June, pp. 228–234.

Kenull, T., Kosyna, G., Thamsen, P. U. (1997): Diagnostics of Submersible Motor Pumps by Non-stationary Signals in Motor Current, *ASME Fluids Engineering Division Summer Meeting*, June, FEDSM97, pp. 1–11.

Martin, C. S. (1983): Representation of Pump Characteristics for Transient Analysis, *ASME Symposium on Performance Characteristics of Hydraulic Turbines and Pumps*, Boston, FED Vol. 6.

Ran, L., Yacamini, R., Smith, K. S. (1996): Torsional Vibrations in Electrical Induction Motor Drives During Start-up, *ASME Journal of Vibration and Acoustics*, April, pp. 242–251.

Shadley, J. R., Wilson, B. L., Dorney, M. S. (1992): Unstable Self-Excitation of Torsional Vibrations in AC Induction Motor Driven Rotational Systems, *ASME Journal of Vibration and Acoustics*, April, pp. 226–231.

Sulzer Brothers Ltd. (1989): *Sulzer Centrifugal Pump Handbook*, Elsevier Applied Science, New York, pp. 59–64, 208–228.

Tsukamoto, H., Yoneda, H., Sagara, K. (1993): The Response of a Centrifugal Pump to Fluctuating Rotational Speed, in *Pumping Machinery*, ASME FED Vol. 154, pp. 293–300.

Vance, J. M. (1988): *Rotordynamics of Turbomachinery*, Wiley, New York.

Wylie, E. B., Streeter, V. L. (1993): *Fluid Transients in Systems*, Prentice Hall, Upper Saddle River, N.J.

6

GENERAL PUMP THEORY

PUMP PERFORMANCE

Pumps use shaft power to increase the energy, pressure, or head of the fluid. The flow rate Q, head H, and efficiency η define the overall performance of a pump. The pump head H corresponds to the increase in total pressure, from inlet flange to exit flange, divided by the specific weight of the fluid, ρg:

$$H = \frac{p_{02} - p_{01}}{\rho g}$$

In the English system of units (lbf, ft, sec) assuming water ($\rho g = 62.4\,\text{lbf}/\text{ft}^3$):

$$H(\text{ft}) = \frac{p(\text{psi})(144\,\text{in}^2/\text{ft}^2)}{62.4\,\text{lbf}/\text{ft}^3}$$

$$Q(\text{ft}^3/\text{sec}) = \frac{Q(\text{gpm})(231\,\text{in}^3/\text{gal})}{(60\,\text{sec}/\text{min})(1728\,\text{in}^3/\text{ft}^3)}$$

The net hydraulic power P delivered by the pump can be calculated from

$$P(\text{hp}) = P(\text{kW})(1.333\,\text{hp}/\text{kW})$$
$$= \frac{\rho g(\text{lbf}/\text{ft}^3) H(\text{ft}) Q(\text{ft}^3/\text{sec})}{550(\text{ft-lbf}/\text{sec})/\text{hp}}$$

Shaft power is calculated by dividing the net hydraulic power P above by the efficiency η.

In the SI system of units (N, m, s, and Pa) assuming water ($\rho g = 9806 \text{ N/m}^3$):

$$H(\text{m}) = \frac{p(\text{N/m}^2)}{9806 \text{ N/m}^3}$$

$$P(\text{hp}) = P(\text{kW})(1.333 \text{ hp/kW})$$
$$= \rho g(\text{N/m}^3) H(\text{m}) Q(\text{m}^3/\text{s})(0.001333 \text{ hp/W})$$

since $1 \text{ N} \cdot \text{m/s} = 1 \text{J/s} = 1 \text{ W}$.

The pump head varies as the flow rate changes. Commercial catalogs show pump performance by means of a plot of the head as a function of the flow rate for a certain shaft speed. The efficiency—the ratio of the output and input power, also a function of the flow rate—also appears on this plot (Fig. 6.1).

Catalog plots often show the head–flow curves corresponding to the same pump with impellers having different diameters. Since the pump head strongly depends on the impeller diameter, machining or trimming the impeller to a smaller diameter certainly offers a simple, fast, expedient way to match the pump characteristics to the system requirement. However, sound design principles show that the flow pattern in such truncated impellers will certainly not be ideal. Chances are that the blade tips will be heavily loaded, and relatively high pressure and flow fluctuations would be expected at the impeller exit. Off-design operation may also be affected, especially in mixed flow pumps, depending on how the impeller exit is modified, since exit recirculation may appear and the head–flow curve may change at reduced flow rates. These effects could be foreseen if a more detailed study of the impeller flow were made.

The nominal rating of a pump corresponds to the head and flow rate at the best efficiency point. These are the conditions for which the pump was designed. It is expected that the flow through the pump would be most disturbance-free at these conditions. Depending on the particular application, pump users desire pumps

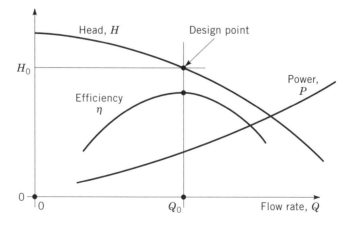

Figure 6.1 Typical pump head–flow curve at constant speed.

OFF-DESIGN OPERATION

Although selected or designed for a particular flow rate, in practice, pumps operate across a certain range of flow rates. At off-design flow rates the flow pattern deviates from the ideal. The flow separates at the impeller blade tips, in the impeller, and in the diffuser or volute. Recirculation may appear at the inlet, flow returning from the impeller eye into the inlet pipe. Recirculation may also occur at the impeller exit, fluid flowing back into the impeller from the diffuser or volute. Not only does the efficiency deteriorate at off-design conditions, but cavitation can appear, which, as will be seen, can destroy the impeller. Excessive vibration appears, which reduces bearing life and reliability, and even the entire pumping system can enter into resonant oscillations—into surging. Some of these phenomena are considered in detail below. Catastrophic failures are unusual, but off-design point operation for a longer period of time seriously reduces pump life.

In selecting a pump, much care must be given in estimating the margin of excess capacity that might be required by the system, supplied by the pump. A much too large margin normally causes the pump to operate far off design, at a bad efficiency value and at conditions that reduce its reliability and life. It becomes an economic consideration whether low efficiency at nominal operating conditions and corresponding operating cost will balance avoiding loss from lack of excess capacity at unusual occasions. Other approaches, such as the standby capacity from a parallel-connected auxiliary pump, for example, might offer a more cost-effective installation. Manufacturers usually provide guidelines for the safe operating range of their pumps (Cooper 1988, Gopalakrishnan 1988).

PUMP TYPES AND SPECIFIC SPEED

Similarity relations can be very useful to scale pumps in size or speed. The performance of very large pumps can be ascertained from testing their geometrically similar, small-scale model in the laboratory. If the test facility cannot provide sufficient suction head or NPSHA, cavitation during testing can be avoided by using low speed and then correcting the data for higher speed. Performance expressed in terms of dimensionless variables helps in selecting optimal configurations (Balje 1962, Cartwright 1977, Tuzson 1993).

The flow rate Q increases in proportion to the shaft speed N, and the head H increases as the square of the speed. A pump rotating twice as fast will deliver twice as much flow and four times as much head. The following relationships can be written in terms of the dimensions of length L and time T, provided that the gravitational acceleration g is also introduced.

$$\frac{Q}{N} = \frac{Q(L^3/T)}{N(1/T)} = \frac{Q}{N}L^3$$

$$\frac{gH}{N^2} = \frac{g(L/T^2)H(L)}{N^2(1/T^2)} = \frac{gH}{N^2}L^2$$

The dimension of the first ratio, representing the flow rate, is L^3; that of the second ratio, representing the head, is L^2. A dimensionless number, expressing the magnitude of the flow rate as compared with the head, is obtained if the first group raised to the second power is divided by the second group raised to the third power:

$$\frac{(Q/N)^2}{(gH/N^2)^3} = \frac{N^4 Q^2}{g^3 H^3}$$

$$= \left(\frac{NQ^{1/2}}{g^{3/4} H^{3/4}}\right)^4 = n_s^4$$

$$n_s = \frac{NQ^{1/2}}{g^{3/4} H^{3/4}}$$

The symbol n_s stands for the dimensionless specific speed which characterizes a pump configuration that is best suited to deliver a certain head H at a certain flow rate Q when operated at a certain speed N. The expression remains dimensionless if consistent units are used, such as $\omega(1/\text{sec}) = N(\text{rpm})(2\pi/60 \text{ sec})$, $Q(\text{ft}^3/\text{sec})$, $g(32.2 \text{ ft/sec}^2)$, and $H(\text{ft})$, for example. The same dimensionless number will be obtained with metric units $\omega(1/\text{sec})$, $Q(\text{m}^3/\text{s})$, $g(9.81 \text{ m/s}^2)$, and $H(\text{m})$.

In U.S. industrial practice a form of the specific speed is used which is not dimensionless but serves the same purpose:

$$N_s = \frac{N(\text{rpm}) Q^{1/2}(\text{gpm}^{1/2})}{H^{3/4}(\text{ft}^{3/4})}$$

$$n_s = 0.0003657 N_s$$

It will be observed that low-specific-speed pumps will have high head and low flow, and high-specific-speed pumps will have low head and high flow rate. Since certain flow losses in pumps depend on the flow rate and others on the head, pump configurations are designed to minimize one type of loss rather than the other, depending on the specific speed. The impellers of low-specific-speed pumps have a large diameter and narrow flow passages, and those of high-specific-speed pumps have large flow passages and relatively small diameter. To maximize the available flow cross section, flow passages in high-specific-speed impellers are inclined from the radial toward the axial direction. These are called mixed flow pumps, in distinction to radial flow pumps.

Past pump designers established by experience typical pump efficiencies as a function of specific speed and flow rate. The plot showing this relationship is shown in Chapter 11, where as a first approximation, an estimated efficiency is required. Pumps of a specific speed of approximately $N_s = 2300$ reach the best efficiency. The corresponding dimensionless specific speed is $n_s = 0.84$. Efficiencies estimated on the basis of industrywide surveys offer a target for pump designers (Sulzer 1989, Turton 1994). For example, the Hydraulic Institute (1994) offers a publication, *Efficiency Prediction Method for Centrifugal Pumps*. Pump technology is old, and excellent designs have evolved over close to 100 years. Efficiencies lower than these targets indicate poor designs, which must have evident shortcomings. Efficiencies higher than these targets testify to outstanding designs, which cannot easily be achieved but require extensive and lengthy analytical and experimental development.

Geometrical scaling of pumps is a powerful method of development. Often, small-scale testing verifies the performance of very large pumps or turbines which cannot be tested in the laboratory. However, small-scale testing raises the question of proper scaling of the efficiency. Scaling affects primarily wall friction, disk friction, and leakage losses, which depend on the relevant Reynolds number. Proper scaling of the efficiency should adjust the individual losses by taking into account the particular Reynolds number that applies. In the case of wall friction, the hydraulic diameter and relative flow velocity in the impeller enter into the calculation of the Reynolds number; in the case of the disk friction, the impeller diameter and tip speed; in the case of clearances, the clearance width and flow velocity. The corresponding friction or loss coefficients also depend on the surface roughness. The partly empirical efficiency scaling formulas of 15 pump experts are reviewed in an article by Osterwalder and Hippe (1982). A properly adjusted and calibrated performance calculation program can help to execute such scaling calculations. It should be kept in mind that efficiency also depends on factors other than the hydraulic design, such as manufacturing methods, mechanical losses, and leakage.

PUMP CONFIGURATIONS

Whereas the design of pump hydraulics follows the guidelines of specific speed, the general arrangement and the configuration of the housing conforms to the needs of the application. In the most common, single-stage, end suction pump configuration (Fig. 6.2), the flow enters horizontally, through the inlet flange, into the impeller eye. The impeller rotates around a horizontal axis and is either mounted directly on the shaft of the electric drive motor, or the pump and motor, separately skid mounted, are connected with a coupling. Closed or open impellers are distinguished depending on whether the impeller has a shroud or the blades are only fixed to the rear hub. In pumps with open impellers a close clearance is

56 GENERAL PUMP THEORY

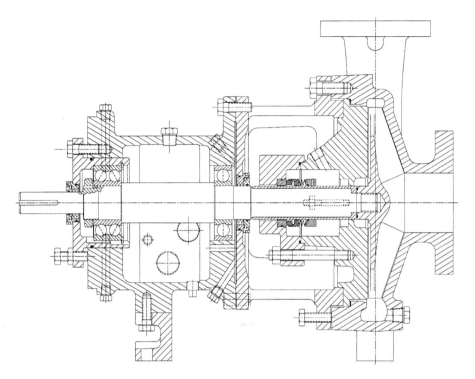

Figure 6.2 Horizontal end suction back pullout (ANSI) pump. (Courtesy of Goulds Pumps, ITT Industries.)

kept between the front of the rotating blades and the stationary housing, and some mechanism exists to adjust this clearance from time to time, since the blade tips may wear. A spiral-shaped volute or collector surrounds the impeller. In process pumps the fluid leaves through an exit flange, with its axis vertical, passing through the centerline, the axis of rotation, of the impeller. These pumps usually carry a designation code consisting of the standard inlet and exit flange sizes. General water pumps often use a long diffuser, aligned off center from the impeller, and tangent to the volute, which allows better diffuser configuration.

Double suction pumps are usually larger and are used primarily in water service (Fig. 6.3). Two back-to-back fused impellers are mounted on a horizontal shaft, supported by bearings on either side. The flow enters through the inlet flange, perpendicular to the direction of the shaft, splits in two, and is ducted to the impeller inlets on either side. A central scroll serves both impellers and leads, through a single diffuser, to an exit flange. Such an arrangement often results in better efficiency because it reduces friction on the back side of the impellers, the disk friction loss, and because by splitting the flow in two, the specific speed of each impeller sometimes becomes more favorable.

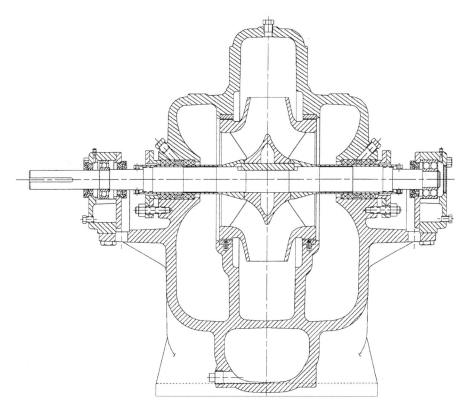

Figure 6.3 Horizontal split-case double suction pump. (Courtesy of Goulds Pumps, ITT Industries.)

The third principal arrangement (vertical pumps or column pumps shown in Fig. 6.4) consists of one or more impellers mounted on a vertical shaft. A vaned diffuser and return passage, leading to the next stage, follows the impeller. The return passage forms part of the housing—the bowl. The fluid enters and leaves the pump in the axial direction. The pump assembly is lowered into a pit. The electric motor, above ground or submerged in the well, drives the pump shaft through a coupling. This arrangement suits water or oil well applications particularly well, because they often require several stages. The well bore diameter limits the impeller size and the pressure rise per stage. The desired pressure is obtained by stacking an appropriate number of standard, mass-produced stages.

A variety of pumps are designed specifically for particular applications, the requirements of which put more emphasis on unusual aspects. Sewage pumps need to pass oversize objects and therefore require exceptionally large flow passages. One-bladed sewage pumps are standard (Stark 1991). Large, slow-

58 GENERAL PUMP THEORY

Figure 6.4 Vertical pump. (Courtesy of Goulds Pumps, ITT Industries.)

running slurry pumps, made of hard metals or rubber, minimize erosion. Multistage boiler feed pumps (Fig. 6.5), nuclear plant coolant pumps, oil well water injection pumps, and certain large process pumps—sometimes driven at high speeds—deliver very high pressures and need high durability and reliability as well as high efficiency. One-of-a-kind large hydroelectric plant pumps—turbines—which one can walk around in are in a class by themselves and need custom design. All these designs derive from the same principles, but require special, advanced development, beyond the scope of this book.

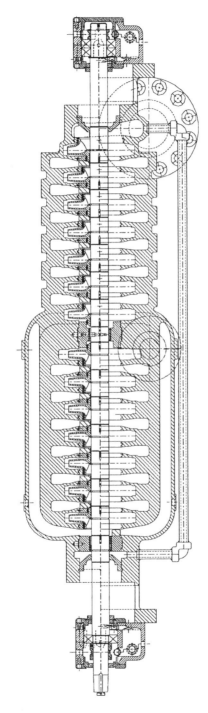

Figure 6.5 Horizontal split-case multistage pump. (Courtesy of Goulds Pumps, ITT Industries.)

REFERENCES

Balje, O. E. (1962): A Study of Design Criteria and Matching of Turbomachines, Part B: Compressor and Pump Performance and Matching of Turbocomponents, *ASME Journal of Engineering for Power*, January, pp. 103–114.

Cartwright, W. G. (1977): Specific Speed as a Measure of Design Point Efficiency and Optimum Geometry for a Class of Compressible Flow Turbomachines, *Scaling for Performance Prediction in Rotodynamic Machines*, Institution of Mechanical Engineers, New York, pp. 139–145.

Cooper, P. (1988): Panel Session on Specifying Minimum Flow, *Proceedings of the Texas A&M 5th International Pump User Conference*, Houston, Texas, pp. 41–47 and 178–192.

Gopalakrishnan, S. (1988): A New Method for Computing Minimum Flow, *Proceedings of the Texas A&M 5th International Pump User Symposium*, Houston, Texas, pp. 41–47.

Hydraulic Institute (1994): *Efficiency Prediction Method for Centrifugal Pumps*, HI, Parsippany, N.J.

Osterwalder, J., Hippe, L. (1982): Studies on Efficiency Scaling Process of Series Pumps, *IAHR Journal of Hydraulic Research*, Vol. 20, No. 2, pp. 175–199.

Stark, M. (1991): Auslegungskriterien für radiale Abwasserpumpenlaufräder mit einer Schaufel und unterschiedlichem Energieverlauf, *VDI Forschungsheft*, Vol. 57, No. 664, pp. 1–56.

Sulzer Brothers Ltd. (1989): *Sulzer Centrifugal Pump Handbook*, Elsevier Applied Science, New York, p. 19.

Turton, R. K. (1994): *Rotodynamic Pump Design*, Cambridge University Press, New York, pp. 10–14.

Tuzson, J. (1993): Evaluation of Novel Fluid Machinery Concepts, *ASME Pumping Machinery Symposium, Fluid Engineering Division Summer Meeting*, Washington, D.C., FED Vol. 154, pp. 383–386.

7

PUMP FLUID MECHANICS

IMPELLER WORK INPUT

The rotating impeller imparts energy to the fluid. It is the most important, the only rotating element of the pump. The diffuser, following the impeller, can transform kinetic energy into pressure energy but cannot increase the total energy of the fluid. The impeller contains radial flow passages, inclined toward the axial direction in mixed flow pumps, formed by rotating blades arranged in a circle. A disk, the shroud, covers the blades on the front. Another disk, the hub, in the back, connects the impeller assembly to the shaft. The flow enters axially near the center of rotation and turns in the radial direction inside the impeller (Fig. 7.1).

Flow conditions, velocities and pressures in the impeller, are described conveniently in terms of cylindrical coordinates: r, θ, and z, in the radial, circumferential, and axial directions. The corresponding three absolute velocity components of the fluid are designated C_r, C_t, and C_z. The three velocities relative to the impeller are W_r, W_t, and W_z. The impeller rotates with an angular velocity of ω, given by $\omega(1/\text{sec}) = N(\text{rpm})(2\pi/60)$.

In mixed flow pumps and in the inlet of radial flow pumps, the streamlines are inclined from the axial direction toward the radial direction. The velocity along projections of streamlines in the r–z plane is called the meridional velocity C_m, which depends on the axial variable z as well as on the radial r. Note that the meridional component of the absolute and the relative velocity are the same, $C_m = W_m$.

The shape of the blades and the resulting flow pattern in the impeller determine how much energy is transferred by a given size impeller and how efficiently it operates (Fig. 7.2). The theoretical energy increase, the theoretical head rise H_{th} through the impeller, can be found by applying the principle of

62 PUMP FLUID MECHANICS

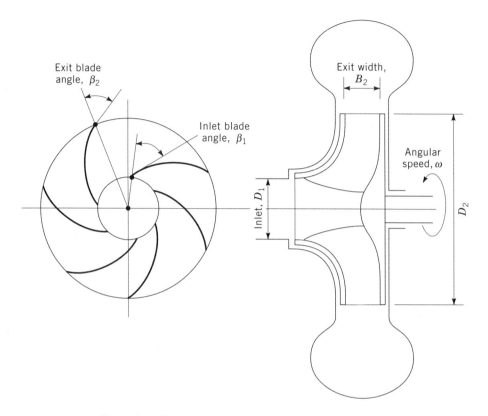

Figure 7.1 Frontal and cross-sectional view of pump impeller.

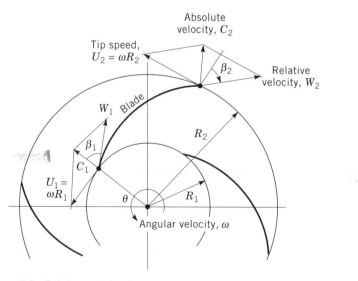

Figure 7.2 Relative and absolute flow velocity vectors in a rotating impeller.

conservation of angular momentum $\rho r C_t$. Considering a control volume of annular shape with an inner radius r_1 and outer radius r_2, the torque T applied to the fluid by the impeller must be equal to the difference between the angular momentum entering, $\rho C_{t1} r_1$, and leaving, $\rho C_{t2} r_2$:

$$\text{torque} = T = Q(\rho C_{t2} r_2 - \rho C_{t1} r_1)$$
$$\text{power} = P = T\omega = \rho Q H_{th} g$$
$$\text{theoretical head} = H_{th} = \frac{(C_{t2} r_2 - C_{t1} r_1)\omega}{g}$$

Applied to pumps, this expression is known as *Euler's equation*. Pumps are usually designed for no angular momentum at the inlet, $C_{t1} r = 0$. Consequently, the equation can be simplified:

$$\text{theoretical head} = H_{th} = \frac{U C_{t2}}{g}$$

where U or $U_2 = \omega r_2$ designates the impeller tip speed, the rotational velocity at the impeller periphery.

FLOW AT IMPELLER INLET AND EXIT

To find the values of the exit angular momentum, the velocities must be examined at the inlet of the impeller, the leading edge of the blades, and at the impeller exit, the trailing edge of the blades. The flow follows certain streamlines inside the rotating impeller, approximately parallel to the blade surfaces. The symbol W designates the corresponding relative velocity, W_1 at the inlet, and W_2 at the outlet. The velocity relative to the impeller can be calculated by adding geometrically the absolute velocity C, measured in the stationary, nonrotating frame of reference, and the rotational velocity of the impeller, $U = \omega r$. Since the velocities are vector quantities, their components in the radial and circumferential direction—W_r, C_r and W_t, C_t, U—must be added separately. Note that the relative radial velocity equals the absolute radial velocity since rotation affects only the tangential components, $W_r = C_r$.

The relative velocity at the inlet W_1 is obtained from the vectorial sum of the absolute velocity approaching the pump inlet C_1 and the rotational velocity $U_1 = \omega r_1$, as shown in Fig. 7.3. Similarly, at the impeller exit the relative velocity W_2 is obtained by the vectorial sum of the absolute velocity C_2 and the tip speed, $U = U_2 = \omega r_2$. The components C_{t2} and C_{r2} in the circumferential and radial directions make up the resultant absolute velocity C_2 since

$$C_2^2 = C_{t2}^2 + C_{r2}^2$$

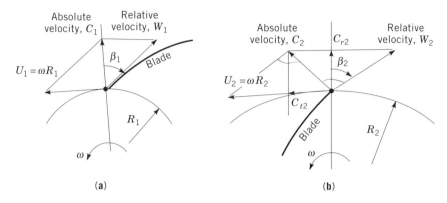

Figure 7.3 Velocity triangles at the impeller inlet (*a*) and exit (*b*).

From the velocity triangle shown in Fig. 7.3 it is apparent that the circumferential component of the absolute velocity can be expressed as a function of the radial velocity component C_{r2} and the local flow angle β_{F2}, which is measured here from the radial direction in the sense opposite to the direction of rotation:

$$C_{t2} = U_2 - W_{t2} = U_2 - W_{r2} \tan \beta_{F2}$$

Consequently, the theoretical head, the energy added to a unit mass of fluid by the pump, can be written

$$H_{th} = \frac{U_2 C_{t2}}{g} = \frac{U_2^2 - U_2 W_{r2} \tan \beta_{F2}}{g}$$

Note that the negative sign appears since the relative velocity W_{r2} points in the direction opposite to the direction of rotation. The blades curve backwards.

As will be seen, this expression is very important for calculating pump performance and for designing new pumps. It assumes that the flow enters the impeller without any swirl, without any circumferential velocity component, C_{t1}, which can usually be assumed in the case of pumps. Strictly speaking, the relationship is valid only for fluid flow on one particular streamline. The velocities and flow angle might vary from one streamline to the next. If the velocity and flow angle are taken as averages on all streamlines at the pump exit, the theoretical head will also be the average head for all streamlines.

SLIP

Unfortunately, the flow does not follow the blades exactly. The flow angle β_{F2} is not identical to the blade angle β_2, because the relative exit velocity W_2 is slightly more inclined opposite to the direction of rotation, as shown in Fig. 7.4. This deviation comes about because the fluid retains its orientation in the absolute

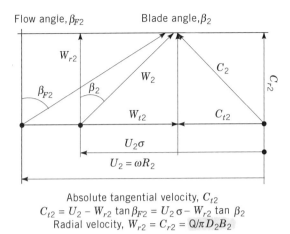

Figure 7.4 Velocity diagram at the impeller exit with slip.

frame of reference, and appears to rotate with respect to the rotating impeller in the opposite direction, which results in a tangential velocity component at the impeller exit opposed to the direction of rotation (Fig. 7.5). To compensate for this deviation a correction is applied to the equation above for the theoretical head in the form of a factor to the tip speed, σ, the slip coefficient:

$$H_{th} = \frac{U_2^2 \sigma - U_2 W_{r2} \tan \beta_2}{g}$$

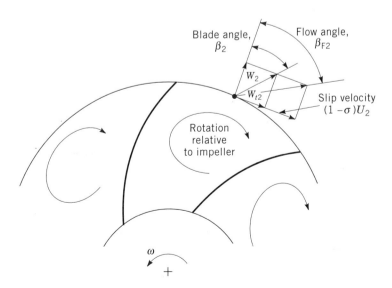

Figure 7.5 Fluid rotation relative to impeller and the resulting slip velocity.

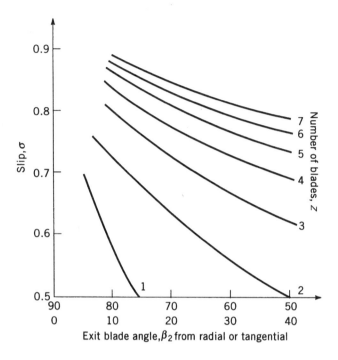

Figure 7.6 Slip coefficient as a function of exit blade angle and number of blades.

Exact values for the slip coefficient have been calculated by Busemann assuming potential flow past logarithmic spiral blades in a radial impeller (Busemann 1928). His results have been verified by modern computer calculations (McDonald and Howard 1973). Wiesner reviewed Busemann's results and the various empirical expressions proposed by several other experts, in addition to a correlation from 49 impellers (Wiesner 1967), and arrived at the following expression for calculating the slip coefficient, which is also shown in Fig. 7.6:

$$\sigma = 1 - \frac{[\sin(90° - \beta_2)]^{1/2}}{Z^{0.70}}$$

In this expression Z stands for the number of blades and β_2 for the exit blade angle measured from the radial or meridional direction in the sense opposite to the sense of rotation. In Wiesner's original paper the impeller exit angle is measured from the tangential direction. This blade angle corresponds to $90° - \beta_2$ if the exit blade angle β_2 is measured from the meridional direction in the direction opposite the sense of rotation, which is the convention adopted in this book. The form of Wiesner's original expression is retained in the expression above to facilitate comparison with his publication. This relationship for the slip coefficient will be used in subsequent chapters.

Slip does not represent an energy loss. It only affects the magnitude of the head that a given size impeller can produce. Its effect is equivalent to the effect of a greater blade exit angle β_2.

RELATIVE VELOCITY IN THE IMPELLER AND ROTHALPY

The inlet velocity triangle defines the inlet relative velocity:

$$W_1^2 = U_1^2 + C_1^2 = \omega^2 r_1^2 + C_1^2$$

Applying Bernoulli's equation to the inlet, we find that

$$h_0 = h_1 + \frac{C_1^2}{2g} = h_1 + \frac{W_1^2}{2g} - \frac{\omega^2 r_1^2}{2g}$$

The last term corresponds to the centrifugal head rise. Consequently, a relationship equivalent to Bernoulli's equation, called the *rothalpy equation*, applies in the rotating impeller in terms of the relative velocity W:

$$\text{rothalpy} = I = h + \frac{W^2}{2g} - \frac{\omega^2 r^2}{2g} - \text{losses}$$

Sometimes it is also called *relative energy* (Wislicenus 1965). This relationship is illustrated in Fig. 7.7. The rothalpy is conserved and remains constant on a streamline through the rotating impeller. This relationship allows calculation of the local static head h, or the pressure, as a function of the radius r anywhere in the impeller, if the relative velocity W on the streamline is known. The relative velocity W would be expected to be calculated from the continuity equation, or approximately from the flow rate, since the flow cross section should be available from the impeller geometry. However, caution is indicated since the relative velocity varies from the pressure side to the suction side of the blades in the circumferential direction due to the blade loading. The pressure side of the blade faces in the direction of rotation. Nonetheless, if a linear variation of the radial velocity W_r can be assumed between the blades in the circumferential direction, the flow rate Q divided by the passage cross section $2\pi r B$ will give the correct magnitude of the average radial velocity. The location of the midstream streamline, however, will be displaced from the center toward the suction side, where the velocity is greater.

From the exit velocity diagram we have the relationship

$$W_2^2 = C_{r2}^2 + (U_2 - C_{t2})^2$$
$$= (C_{r2}^2 + C_{t2}^2) + U_2^2 - 2U_2 C_{t2} = C_2^2 + U_2^2 - 2U_2 C_{t2}$$

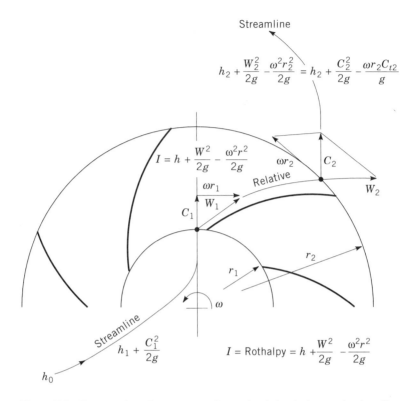

Figure 7.7 Conservation of energy equation and rothalpy in the rotating impeller.

Observing the conservation of rothalpy I, and substituting W_2 yields

$$h_1 + \frac{C_1^2}{2g} = I = h_2 + \frac{W_2^2}{2g} - \frac{U_2^2}{2g} = h_2 + \frac{C_2^2}{2g} - \frac{U_2 C_{t2}}{g}$$

$$\left(h_2 + \frac{C_2^2}{2g}\right) - \left(h_1 + \frac{C_1^2}{2g}\right) = \frac{U_2 C_{t2}}{g} = H_{\text{th}}$$

Therefore, the total head of the flow passing through the impeller, the sum of the static head h and the velocity head $C^2/2g$, increases by the theoretical head, which is added by the impeller, $UC_{t2}/g = H_{\text{th}}$, as stated above.

Although the rothalpy equation, strictly speaking, is valid only along a streamline, in most pumps the total energy at the inlet is the same on all streamlines, and therefore the rothalpy is also the same for all streamlines. Any pressure drop due to frictional losses in the impeller must be deducted from the static pressure. At the impeller exit the flow angle β_{F2} and the relative velocity W_2—therefore, the tangential velocity component C_{t2} also—may vary from one streamline to the next. Consequently, the theoretical head, the energy addition by the pump, may not be the same for all streamlines. The magnitude of the relative velocity W_2 can be affected, for example, if the exit radius r_2 varies, as in mixed flow pumps, with a slanted exit. We consider such a case in Chapter 14. Friction losses can also make the energy addition uneven, as in the case of a jet-wake flow pattern, discussed below.

Since one earmark of a good pump is a uniform theoretical head, or equal-energy addition on all streamlines, the flow variables need to be calculated on several streamlines. Impeller design calculations consist of determining the angle β along the trailing edge of the blade—and more generally along the entire blade surface—so that uniform energy be imparted to the fluid on all streamlines. Usually, calculations follow a streamline along the hub, the shroud, and along the midstream. More detailed calculations also consider streamlines on the pressure and suction sides of the blades. The use of computers allows calculations on multiple streamlines covering the entire flow passage. The choice of computation complexity and time expenditure should correspond to the value of the pump.

As will be seen, at off-design operation most pumps, mixed flow pumps in particular, do not impart the same energy on all streamlines, which results in a shear flow at the impeller exit. This flow pattern leads to flow losses in the form of eddies and flow fluctuations, and can sometimes result in peculiar behavior of the pump.

Since the velocity and pressure vary from the pressure side—the side facing in the direction of rotation—to the suction side of the blades at the impeller exit, a stationary observer looking at the rotating impeller will see a fluctuating, unsteady flow and pressure field sweeping by. If Z number of blades pass per shaft revolution, an oscillation frequency $f = ZN$ per minute or $f = ZN/60$ per second will be observed. This fundamental pressure oscillation, which is present in all pumps to a greater or lesser extent, is called the *blade passing frequency*. It can induce resonant vibrations in the piping system connected to the pump. The magnitude of the pressure fluctuations varies in proportion to the blade loading, the pressure difference between the two sides of the blade, near the impeller exit.

FLOW SEPARATION IN THE IMPELLER

To illustrate flow conditions in the impeller, most designers plot the relative velocity W, along the pressure and the suction side of the blade, against the streamline length, measured from the leading edge, as shown in Fig. 7.8. Because of the blade loading—the pressure difference across the blade—the pressure side velocity is lower than the suction side compared at the same fraction of the streamline length. The rothalpy equation expresses this relationship between the local head or pressure and the velocity. The difference between the velocities on either side of the blade illustrates the pressure difference or blade loading, but only qualitatively. Actually, the rothalpy equation relates the pressure not to the velocity but to the square of the velocity. For a quantitative measure of the pressure difference, the square of the velocities would need to be plotted. The plot shows that if the blade loading becomes too large, the velocity on the pressure side approaches zero. The flow separates, which must be avoided.

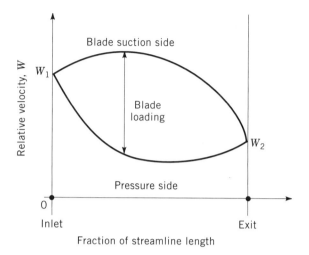

Figure 7.8 Relative velocity difference in the impeller as an approximate measure of blade loading.

An approximate estimate for the appearance of a separated flow can be obtained by applying the principle of the angular momentum change to a very narrow annular volume, such that $r_2 = r_1 + \Delta r$ and having width B in the axial direction, as illustrated in Fig. 7.9. Then the incremental torque ΔT corresponds to the moment of the pressure force exerted on the blades at a radius of r_1 in the circumferential direction. The pressure on the side of the blade facing the direction of rotation, the pressure side p_p must be greater than the pressure acting on the trailing, suction side p_s of the blade. In the case of an impeller having Z number of blades, and if the incremental increase in the angular momentum over a small radial distance of Δr is $\rho(\Delta C_t r_1)$, then

$$p_p - p_s = \frac{\Delta T}{Z \, \Delta r \, B r_1} = \frac{Q}{2\pi r_1 B \, Z} \frac{2\pi \rho \, \Delta C_t \, r_1}{\Delta r}$$
$$= \frac{2\pi}{Z} C_r \frac{\rho \, \Delta C_t \, r_1}{\Delta r}$$

It appears from this equality that the energy added to the fluid is proportional to the pressure difference across the blade, which is usually referred to as the *blade loading*. In the flow passage between two blades, the pressure decreases in the circumferential direction, approximately linearly, from the pressure side of the blade facing in the direction of rotation to the suction side. As will be seen, there is a corresponding increase in the velocity in the impeller. If the blade loading, and therefore the pressure difference, becomes very high, the velocity at the pressure side of the blade can become very small or vanish, which results in flow separation and losses. Generally speaking, the blade loading provides a measure of the likelihood of a flow separation.

FLOW SEPARATION IN THE IMPELLER

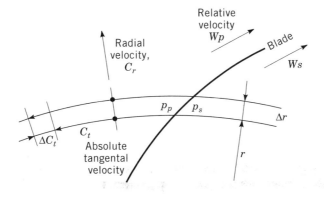

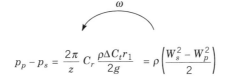

$$p_p - p_s = \frac{2\pi}{z} C_r \frac{\rho \Delta C_t r_1}{2g} = \rho \left(\frac{W_s^2 - W_p^2}{2} \right)$$

Figure 7.9 Pressure difference across the impeller blade: blade loading.

The incremental, radial change in the angular momentum corresponds to the radial derivative of the expression of the angular momentum:

$$rC_t = r^2 \omega \sigma - rC_r \tan \beta$$

Assuming that the radial velocity component C_r and the blade angle β do not change, or change very slowly, in the radial direction, and that the streamlines are parallel to the blades in the circumferential direction, the incremental, radial change of the angular momentum becomes

$$\frac{\Delta(rC_t)}{\Delta r} = 2r\omega\sigma - C_r \tan \beta = 2U\sigma - C_r \tan \beta$$

Substituting into the expression for the pressure difference across the blade, and remembering that, according to the rothalpy equation, the pressure difference corresponds to a velocity head difference at the same radius, one obtains:

$$\frac{p_p - p_s}{\rho} = \frac{W_s^2 - W_p^2}{2}$$
$$= \frac{(C_{rs}^2 - C_{rp}^2)(1 + \tan^2 \beta)}{2} = \frac{2\pi}{Z} C_{rs}(2U\sigma - C_{rs} \tan \beta)$$

From this expression those conditions can be determined for which the velocity at the pressure side of the blade becomes zero ($C_{rp} = 0$). These conditions signal separation in the impeller and the onset of a jet-wake flow pattern.

$$\frac{C_{rs}}{2} = \frac{(2\pi/Z)(2U\sigma)}{1 + \tan^2 \beta + 2(2\pi/Z)\tan \beta}$$

When the average radial velocity $Q/(2\pi rB)$ becomes smaller than the value of the radial velocity $C_{rs}/2$ above, separation appears. For radial blades, when $\tan \beta = 0$, the radial velocity would have to be on the order of the circumferential speed U to avoid separation, which would be excessive, and separation would be expected unless the number of blades Z were large. Indeed, in centrifugal compressors, with blades nearly radial, 30 blades are not unusual. In centrifugal pumps, with typically about six blades ($Z = 6$) and a blade angle of about $68°$ ($\beta = 68°$), the radial velocity would have to be less than about 15% of the circumferential speed U for separation to appear, according to the criterion above, which would rarely be the case in practice at design conditions.

Good pump design also demands that blade loading, and the corresponding velocity difference between the pressure and suction sides, gradually grow at the inlet and taper off at the exit. Consequently, some designers use the plot of Fig. 7.8 to verify a suitable velocity distribution and satisfactory impeller design (Dallenbach 1961).

The danger of flow separation is also present when the velocity decreases and the pressure increases too rapidly on any streamline. As discussed above, the velocity cannot change beyond the separation point, and no additional velocity head can be converted to pressure head. In centrifugal pumps the inlet relative velocity W_1 is usually larger than the exit velocity W_2, and therefore a certain diffusion—slowing of the velocity—must take place in the impeller, which should not be excessive. Empirical rules exist for a tolerable velocity ratio, usually on the order of $W_2/W_1 = 0.7$. Separation and increasing flow losses would then be present for velocity ratios below this value. The first flow separation in the impeller usually appears on the shroud and on the suction side of the blades. The highest inlet velocity occurs at this location, and the greatest diffusion or deceleration takes place. A sharp curvature of the shroud in the radial plane contributes to the likelihood of separation, and should be avoided.

Conditions become necessarily critical at some point with decreasing flow rate, since the inlet velocity W_1 approaches the inlet circumferencial velocity U_1, while the exit velocity W_2 approaches zero as the flow rate is reduced. The ratio of velocities W_2/W_1 decreases toward zero. Therefore, flow separation necessarily appears at some point as the flow rate is reduced below the best efficiency point. Such a separated flow can trigger, or is replaced by, inlet recirculation, which is discussed in Chapter 14.

JET-WAKE FLOW PATTERN

It was pointed out above that a shear layer, or separation streamline, between two flow regions of different fluid energy is stabilized if the acceleration and corresponding pressure rise perpendicular to the shear layer are directed from the low-energy region toward the high-energy region. When separation appears in the impeller, two flow regions of different energy can be distinguished: the separated region and the main stream, as shown in Fig. 7.10. The relative velocity W remains constant, but different, on either side of the streamline separating the two regions. The acceleration, and corresponding pressure increase, are directed from the suction side to the pressure side. A separated region of low energy will be stable and can then persist on the suction side. A low-energy separated region—for example, a boundary layer—on the pressure side would be unstable and would tend to migrate along the hub and shroud surfaces to the suction side. Such secondary flows do exist and have been observed in impellers (Bayly and Orszag 1988; Johnston et al. 1972, Johnston and Rothe 1976; Tuzson 1977). They are discussed in Chapter 8.

In the case of heavy blade loading, potential flow calculations may predict that the velocity slows to zero at the pressure side and that flow separation will occur on the pressure side. However, since a separated region cannot persist on the pressure side, the flow switches to a jet-wake flow pattern. Consequently the ideal potential flow model gives a false picture of the flow pattern. A flow pattern consisting of a stable separated region or wake on the suction side and a main stream or jet near the pressure side of the blades is called a *jet-wake flow pattern*. Such a flow pattern appears only under certain conditions (Dean and Senoo 1960, Tuzson 1993).

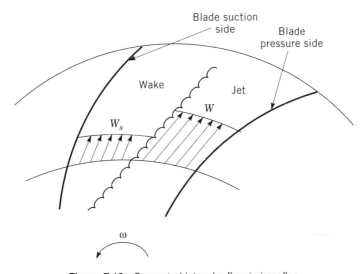

Figure 7.10 Separated jet-wake flow in impeller.

74 PUMP FLUID MECHANICS

In the separated region a static head h_s and a velocity W_s are assumed. The velocity W_s is smaller than the free-stream velocity W, the velocity head difference, $(W^2 - W_s^2)/2g$, having been lost. The rothalpy equation still applies separately in each region, except that the rothalpy I_s in the separated region is smaller by the amount of the foregoing head loss. However, the loss can be compensated if the energy addition in the separated region is higher than in the free stream.

If β is the flow angle of the separation streamline measured from the radial in the direction opposite to the sense of rotation, the energy addition in each region is given by

$$\frac{UC_t}{g} = \frac{U^2 - UW\sin\beta}{g}$$

$$\frac{UC_{ts}}{g} = \frac{U^2 - UW_s\sin\beta}{g}$$

The energy addition exactly balances the loss when

$$\frac{W^2 - W_s^2}{2g} = \frac{U(W - W_s)(\sin\beta)}{g}$$

$$\frac{W + W_s}{2U} = \sin\beta$$

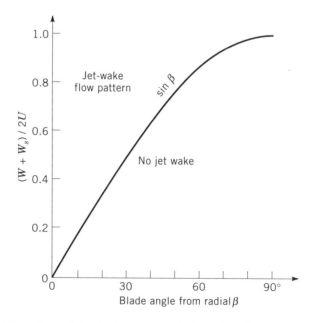

Figure 7.11 Jet-wake flow pattern criterion. (From Tuzson 1993.)

Therefore, a stable separation on the suction side and a jet-wake flow pattern can persist if the ratio of the average velocity on the separating streamline, $(W + W_s)/2$, and the circumferential speed U is larger than $\sin \beta$. The flow angle β can be taken equal to the local blade angle. Figure 7.11 illustrates this relationship. It is apparent, for example, that radial bladed impellers, when $\beta = 0$, would usually be expected to operate with a jet-wake flow pattern. The blades in most pumps lean strongly backward, making $\sin \beta$ relatively large. Consequently, jet-wake flow pattern would not be expected in pumps.

BLADE-WAKE MIXING

Although usually thin, impeller blades have a definite thickness and leave a wake in the flow at the impeller exit. It can be assumed that the cross section of the flow passage between the blades of the rotating impeller suddenly increases by the amount of the blade thickness as the flow leaves the impeller. Since the blade thickness is considerably less than the passage width, the area increase and the corresponding loss are relatively small and may or may not be negligible. A sudden expansion loss calculation can provide a rough estimate of the magnitude of such a blade-wake mixing loss. There is reason to believe, however, that such a flow model is not valid and that the mixing loss is relatively smaller than would be expected.

In reality the flow leaving the impeller is unsteady when considered by a stationary observer in an absolute frame of reference. The absolute flow velocity at the impeller exit is not parallel to the wake. The mixing process is therefore entirely different from a steady wake mixing. The fluid in the wake is not low-energy fluid as in the case of a wake behind a stationary object, but can move with a higher velocity than the main flow, approximately with the tip speed. In addition, trailing vortices are likely to be present in the corners between the blade and the hub and shroud surfaces.

Very fast mixing has been observed in tests, causing surprisingly low losses. The explanation was offered that energy transfer takes place by normal, unsteady pressure forces, by pressure exchange (Foa 1958, 1973) rather than by viscous or turbulent shear forces. Pressure exchange is considerably more efficient than shear in transferring energy. Conceptually, the issue is very important, but since the losses are low and mixing is fast, it does not affect pump design or pump performance significantly. However, blade-wake mixing would be expected to leave a high level of turbulence in the flow, not to mention the trailing vortices.

VANED DIFFUSER

In some pumps a radial, vaned diffuser follows the impeller exit. A short radial distance, a vaneless space, precedes the diffuser. The diameter of the diffuser inlet is usually about 5 to 10 % greater than the impeller diameter. As described above, the pressure and relative velocity vary from the pressure side of the blades to the suction side at the impeller exit. In the stationary frame of reference a fluctuating flow comes off the impeller. If the leading edge of the diffuser vanes is too close to the impeller, the flow fluctuations impinge on the diffuser vanes and can produce undesirable noise and pressure pulsations at the pump exit. The vaneless space serves to allow the flow to mix, become more uniform, and adjust to the inlet of the vaned diffuser. The pressure pulsations can be reduced by slanting the leading edge of the diffuser vanes in the circumferential direction with respect to the trailing edge of the impeller blades. The trailing edge of the impeller blades then sweeps gradually across the leading edge of the diffuser vanes instead of overlapping exactly at a certain moment. On small multistage pumps sometimes a few widely spaced diffuser vanes strongly inclined in the tangential direction are positioned relatively close to the impeller exit. Their purpose is to attempt to slow down the high-speed flow immediately after the impeller exit and recover some pressure. The radial component of velocity in the diffuser is imposed by the continuity equation, the ratio of the flow rate, and the available cross-sectional area. Diffusion is accomplished primarily by reducing the tangential component of the velocity.

The direction of the diffuser leading edge and the diffuser throat size, the cross-sectional area between two vanes perpendicular to the velocity, must match the magnitude and direction of the velocity approaching from the impeller exit. The velocity may slow down somewhat in the radial vaneless space in proportion to the increased radius: $C_3 = C_2(R_2/R_3)$ while maintaining its direction. On the other hand, the diffuser throat velocity C_{Q3}, assuming straight vanes and sidewalls, is given by the expression

$$C_{Q3} = \frac{Q}{A_s} = \frac{Q}{2\pi R_3 B_3 \cos \beta_3} = C_3 = C_2 \frac{R_2}{R_3}$$

where A_s is the diffuser throat area, R_3 the radius at the diffuser inlet, B_3 the width of the diffuser, and β_3 the diffuser inlet angle. The expected pressure recovery in the vaned diffuser follows the traditional diffuser experience. The effectiveness of diffusers consisting of widely spaced airfoil-shaped vanes is more difficult to estimate. General airfoil theory might provide a measure of the turning of the velocity, and corresponding diffusion, that could be expected.

Multistage pumps combine the vaned diffuser with return-flow passages, which conduct the flow to the inlet of the next stage. The purpose of the return passages is to turn the swirling flow in the axial direction. They don't necessarily slow down the fluid. In some pumps the impeller absolute exit velocity is approximately equal to the inlet velocity. Under such conditions no opportunity exists to slow down the flow and recover pressure. No diffuser is needed. In return passages only limited pressure recovery can be expected, concentrated near the inlet section of the diffuser. A complex three-dimensional flow is generated in the subsequent return bend, which is not suitable to slow the flow and recover pressure. The flow velocity in the return bend is usually kept constant and is turned gradually in the axial direction.

At off-design flow rates the impeller exit velocity is not aligned with the diffuser inlet, and leading edge separation occurs. Also, a mismatch appears between the approach velocity and the throat velocity, which results in a loss. Less head is lost if the flow has to accelerate into the diffuser throat than if the flow has to slow down. The flow has to accelerate at flow rates above the design flow rate and has to slow down at flow rates below the design flow rate. Theoretically, the diffuser should be correctly matched at design flow conditions.

VOLUTE

The impeller often discharges directly into the volute, which is a spiral-shaped flow passage usually of circular or trapezoidal cross section (Hagelstein et al. 1999) (Fig. 7.12). Its cross section increases gradually around the impeller periphery, starting from the volute tongue and ending at the volute throat. The volute tongue directs the total flow, collected from around the impeller, through the throat to the pump exit flange. A diffuser may or may not exist between the

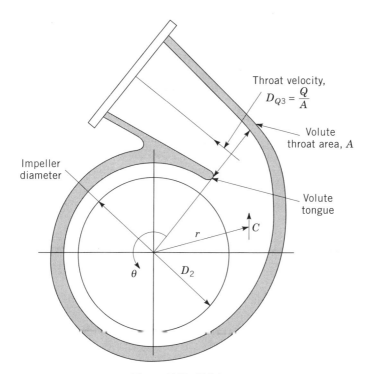

Figure 7.12 Volute.

volute throat and the exit flange, depending on the available space and the flow velocities.

The gradually increasing volute flow cross sections are calculated from the flow rate and from an average velocity at the volute cross section center. The usual flow model assumes that the impeller exit tangential velocity decreases in proportion to the radius to maintain constant angular momentum. The volute cross section A as a function of the circumferential angle θ counted from the tongue becomes

$$A = \frac{(\theta/2\pi)Q}{(C_{t2}R_2/r)}$$

This expression also determines the needed volute throat cross section.

When the flow rate is greater than the design flow rate, the flow generally accelerates in the volute, and the pressure tends to decrease in the circumferential direction. When the flow rate is less than the design flow rate, the volute velocity tends to decrease, and the pressure increases circumferentially around the impeller. Because of this pressure increase or decrease, a transverse pressure force appears on the impeller at off-design conditions, which must be resisted by the pump bearings (Agostinelli et al. 1960, Csanady 1962, Lorett and Gopalakrishnan 1986, Iversen et al. 1960, Schwarz and Wesche 1978).

The static pressure distribution at the volute periphery can easily be checked experimentally because the volute castings are usually provided with threaded draining holes, at which pressure measuring instruments can be attached. As in the case of the vaned diffuser, a radial space is left between the impeller periphery and the volute tongue. The circle centered on the center of rotation and touching the volute tongue, the base circle, usually has a diameter 10 % greater than that of the impeller. At flow rates less than the design flow rate, the volute tongue deflects some of the flow approaching the volute throat, which passes between the volute tongue and the impeller and returns into the volute. If the space between the tongue and the impeller is very large, too much flow returns unnecessarily, and losses increase. If the space is too small, strong pressure fluctuations at the blade passing frequency can occur.

In some self-priming pumps a hole is drilled from beyond the volute throat across the tongue back into the volute, to recirculate the fluid until the pump is fully primed. The detailed geometry of the volute and the size and location of the hole are usually determined by experiment. They indicate the amount of suction lift the pump can handle during priming.

INCIDENCE

The flow separates at the leading edge of the impeller blade if the direction of the relative velocity W_1 does not align with the leading edge, as shown in Fig. 7.13. Correct pump design usually assures correct alignment between the velocity

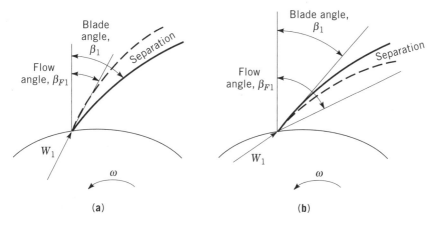

Figure 7.13 Inlet blade incidence: (a) increased flow rate, $Q > Q_0$; (b) reduced flow rate, $Q < Q_0$.

vector and the blade at the inlet at the best efficiency flow rate. At flow rates greater than design flow rate the flow angle β_{F1}—measured from the meridional or radial direction—is smaller than the blade angle at the inlet β_1, and a separated flow region appears on the pressure side of the blade. At flow rates below design flow rate, separation occurs on the suction side of the blade. The term *incidence* designates the difference between the flow angle and the blade angle: $\beta_{F1} - \beta_1$. The separated region on the pressure side of the blade becomes unstable. Flow oscillations and rapid mixing ensue. Separation on the suction side remains stable and can persist far into the impeller.

The blockage from leading-edge separation restricts the flow passage cross section and therefore increases the local velocity. The head loss incurred when the separated region mixes with the main stream can be considered a sudden expansion loss and can be calculated from the velocities at separation and downstream, where complete mixing has taken place (Cornell 1962). The corresponding loss calculation procedure is presented in Chapter 10.

Particular attention needs to be paid to calculation of the blade angles at the leading edge. Since the inlet flow of centrifugal pumps turns from the axial direction to the radial direction, the relative flow velocity W_1 and its direction vary along the leading edge. The flow pattern is three-dimensional, and a three-dimensional flow calculation would be desirable. A simplified calculation procedure is presented below and used in the INLET computer program in the Appendix.

The real flow conditions at the impeller blade inlet at incidence are often much more complicated than those of the two-dimensional separation model. If the flow velocity meets the leading edge at a slant, a tip vortex can develop in the separated region along the leading edge, which trails away toward the hub (Fig. 7.14). High-specific-speed pumps are particularly sensitive to inlet conditions because the through flow velocities are relatively high and the inlet diameter is

80 PUMP FLUID MECHANICS

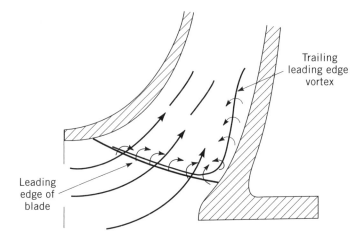

Figure 7.14 Incidence and separation on slanted leading edge: leading-edge vortex.

also relatively large. The circumferential velocity on the shroud far exceeds the circumferential velocity at the hub. Meridional velocities are high on the shroud and low on the hub because of the curvature of the streamlines. Since flow losses vary with the square of the velocity, high velocities indicate potentially high losses. The most critical conditions are encountered on the shroud streamline.

CAVITATION

Chapter 9 is devoted to the subject of cavitation in pumps. Only a brief overview is presented here.

The high local velocity, near the blade leading edge, corresponds to low static pressure, and cavitation could appear. Under cavitating conditions the absolute local static pressure falls below the vapor pressure of the fluid, and steam or gas bubbles appear. The evolving gas, and corresponding volume increase, can choke down the flow and prevent further increase of the flow rate. The head–flow characteristics of the pump then end at this maximum flow rate.

The collapse of the cavitation bubbles, and the reentering fluid jet, impinging on the blade surface, can generate very high local pressures and can destroy the blade material. Such cavitation damage must be avoided and remains a critical issue in pump design, selection, and operation. Manufacturers' catalogs provide information on the capability of the pump to tolerate low inlet pressures. Plots usually show the net positive suction head required (NPSHR) by the pump as a function of flow rate. In cavitation test the pressure at the inlet flange of the tested pump is lowered until the head delivered by the pump drops off by 3 %, due to cavitation. NPSHR is defined as the absolute total pressure at the impeller inlet

p_{01}, less the vapor pressure of the fluid p_{vp}, divided by the specific weight of the fluid:

$$\text{NPSHR} = \frac{p_{01} - p_{vp}}{\rho g} = h_{01} - h_{vp}$$

Empirical data were compiled by the Hydraulic Institute on the NPSHR requirement of typical pumps, which can be summarized by the expression

$$\text{NPSHR} = 0.415 \times 10^{-6} N_s^{5/3} H - h_{vp} = 0.415 \times 10^{-6} N_s^{1/3} N^{4/3} Q^{2/3} - h_{vp}$$

Here h_{vp} is the head corresponding to the vapor pressure of the fluid, about 1 ft in terms of absolute pressure head for water at standard conditions. The specific speed N_s is in English units, Q (gpm), N (rpm), and H (ft).

Industry standards (ASME, API, ISO, Hydraulic Institute) prescribe test procedures for the experimental determination of NPSHR. These generally base the definition of NPSHR on that pressure head at the pump inlet at which the pump head drops off by 3 %. However, cavitation noise and cavitation damage usually precede the head drop as the suction pressure decreases.

In Chapter 9 we present analytical models rather than empirical correlations for estimating the conditions for the onset of cavitation in pumps. The analytical models lead to the calculation of the local absolute static pressures near the blade leading edges, where cavitation is most likely to appear first. Cavitation is expected to appear when the local absolute static pressure reaches the vapor pressure of the fluid. Knowing the flow conditions at the pump inlet, when cavitation sets in, an analytical expression for estimating the NPSHR required by the pump can be derived from fluid mechanic fundamentals.

When specifying or matching pumps to a certain application, the net positive suction head available (NPSHA) from the system is calculated by estimating the total absolute pressure available from the system at the pump inlet, p_{01}. The NPSHA then becomes

$$\text{NPSHA} = \frac{p_{01} - p_{vp}}{\rho g} = h_{01} - h_{vp}$$

This value of NPSHA is then compared with the value of NPSHR from the pump manufacturer's catalog. The NPSHA values must be greater than the NPSHR values.

In practice, the appearance of cavitation or vapor bubbles depends on a variety of factors. Absorbed or dissolved air, gas, or fine bubbles, for example, may exist in the fluid, which evolve or expand with decreasing pressure long before the vapor pressure of the fluid would be reached and boiling could start. Symptoms of cavitation and malfunctioning of the pump can already appear at pressures above the vapor pressure of the fluid. Such malfunctioning may occur, for example, when the inlet piping is not tight. Because of the negative pressure in the inlet, air can then be aspirated into the inlet through the leakage paths. Therefore, the NPSHA values available from the system should be generally higher than the NPSHR value required by the pump as listed by the manufacturer. The temperature of the fluid affects the vapor pressure and may also lead to premature cavitation. Hot or boiling fluids may require pressurization of the pump inlet. Fluids other than water have vapor pressures that require a different NPSHR value than that needed with water.

DISK FRICTION ON THE IMPELLER

Practical hydraulic pump design must consider all hydraulic forces on the impeller. The circumferential, fluid frictional force on the impeller, disk friction, absorbs shaft power and affects the overall efficiency of the pump. The calculation procedure for estimating the corresponding power loss is given in Chapter 10.

Clearance spaces must exist between the stationary housing and the rotating hub and shroud faces of the impeller. In centrifugal pumps the external faces of the impeller on the hub and shroud side resemble a disk that is rotating in a stagnant fluid (Fig. 7.15). The fluid at the disk surface rotates with the circumferential velocity of the disk. The fluid shear at the rotating disk tends to accelerate the fluid volume in the clearance, whereas shear at the stationary housing wall tends to slow it down. Consequently, the fluid volume in the clearance between the impeller face and the housing will rotate with a velocity about half that of the circumferential velocity of the impeller. Since the disk friction torque increases with the fifth power of the diameter, small changes in the

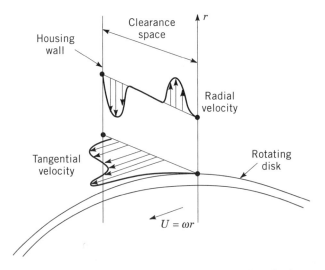

Figure 7.15 Flow velocity profiles in clearance between rotating impeller face and housing.

diameter have a disproportionately large effect on the disk friction torque. The outer periphery of the disk contributes substantially to the total friction force.

The circumferential velocity of the fluid in the clearance changes rapidly near the disk, perpendicular to the surface, over a certain, small distance, the boundary layer. A similar boundary layer exists on the housing wall. If the clearance between the impeller and the housing is much greater than the thickness of these boundary layers, the clearance width does not influence the magnitude of the friction force, the disk friction, on the impeller. While the boundary layer thickness, and more generally the flow pattern in the clearance space, can be affected by the leakage flow through the clearance space, an approximate value of the boundary layer thickness can be obtained. It can be estimated to be a low multiple of $(\nu/\omega)^{1/2}$, which in water at 1800 rpm amounts to a few thousands of an inch (Schlichting 1960). The disk friction losses of the pump will be minimized if the clearance width, between the impeller faces and the housing walls is made much larger than the boundary layer thicknesses.

The fluid rotating near the impeller face generates a radial pressure gradient because of the centrifugal acceleration. This radial acceleration induces a radial, secondary flow outward. A corresponding radial flow exists near the housing wall, which is directed inward. Therefore, fluid circulates radially out along the impeller face and radially in along the housing wall. In low-specific-speed pumps producing high head and low flow rate, the secondary flow flowing radially out along the impeller face can become significant compared with the flow rate of the pump. Since this secondary flow rotates with the circumferential speed of the impeller, it carries high energy. It was suggested that on small pumps, a portion of the head of this secondary stream should be added to the head produced by the pump (Hamkins 1984). The outside faces of the impeller would then act as separate pumps contributing to the output of the flow inside the impeller.

Special pump configurations have been proposed and tested, the impellers of which consist of closely spaced parallel disks mounted on the shaft. The fluid is entrained by the fluid friction on the disks without blades (Hasinger and Kehst 1963, Roddy et al. 1987). Such disk pumps found special applications with heavily contaminated and viscous fluids, for example.

If considerable leakage flow passes through the clearance space between impeller and housing, it can cause a change in the average rotational velocity of the fluid volume in the clearance (Rohatgi and Reshotko 1974).

In the turbulent flow regime, which is the normal operating range of centrifugal pumps, the shear force per unit surface area of the disk must be proportional to the density ρ and to the square of the circumferential velocity of the disk, $\omega^2 r^2$. The surface area of the disk is approximately proportional to the square of its radius R^2. The moment of this force, M, with a moment arm of R, produces a resisting torque on the pump shaft, which will be given by an expression of the form

$$M = C_m \rho \omega^2 R^5$$

By substituting consistent units, it can be verified that the coefficient C_m is dimensionless. Experiments have shown that, if moment M represents the resisting torque on both sides of a disk, the coefficient C_m takes on values from 0.006 to 0.01 with increasing roughness of the disk surface (Nece and Daily 1960). In practice (as will be shown), an appropriate value of the coefficient C_m should be established from matching the head–flow curve of an existing, typical pump with a performance calculation program. Protruding bolts and unevenness will affect the coefficient (Zimmermann et al. 1986). Operating on a fluid other than water will also require the use of a different coefficient, since the Reynolds number will be affected by the fluid viscosity.

AXIAL THRUST ON THE IMPELLER

The rotating fluid volume in the clearance between the impeller and the housing results in a radially increasing pressure distribution. A pressure force is exerted on the front, the shroud, and on the back, the hub, of the impeller. The resultant of these two opposing forces must be resisted by the thrust bearing of the pump and therefore needs to be calculated (Due 1966, Guelich et al. 1989; Fig. 7.16). Assuming that both clearances communicate with the impeller exit, the same impeller outlet pressure also exists in the clearances at the outer radius. The static pressure at the impeller exit p_2, above the static pressure at the inlet p_1, can be calculated either in terms of the theoretical head and the absolute velocities, or in terms of the circumferential and relative velocities if the rothalpy equation is used.

$$p_2 - p_1 = \rho g H_{th} - \frac{\rho(C_2^2 - C_1^2)}{2} = \rho U_2 C_{t2} - \frac{\rho(C_2^2 - C_1^2)}{2}$$

$$= \frac{\rho \omega^2 (R_2^2 - R_1^2)}{2} - \frac{\rho(W_2^2 - W_1^2)}{2}$$

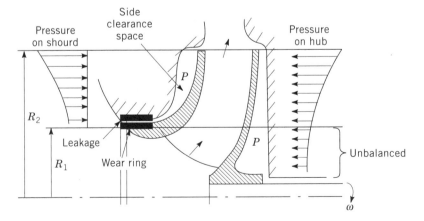

Figure 7.16 Axial pressure forces and thrust on impeller.

It can be assumed that the pressure in the clearance space decreases with decreasing radius in proportion to the centrifugal acceleration, which results from the rotation of the fluid volume in the clearance. The rotational velocity of the fluid in the clearance space is approximately one half that of the impeller, $\frac{1}{2}\omega r$. The pressure difference between the outer and inner radius of the clearance space, which must be deducted from the impeller exit static pressure, becomes

$$\rho(\tfrac{1}{8})\omega^2(R_2^2 - r^2)$$

where R_2 is the outer radius, and r is the radius at which the pressure is evaluated.

The local pressure above the inlet pressure on the impeller face at a radius r becomes

$$p - p_1 = (p_2 - p_1) - (\tfrac{1}{8})\rho\omega^2(R_2^2 - r^2)$$

The inner radius R_1 on the front, the shroud, of the impeller corresponds to the radius of the wear ring or front sealing sleeve. The rear, hub cavity usually extends to the radius of the pump shaft, which has a much smaller radius than that of the wear ring. The pressure force on the rear side of the impeller would then be greater than on the front, and without some special provision, their difference might represent an excessive thrust bearing load. Therefore, on larger pumps, a sealing sleeve is made on the hub also, which has the same radius as the wear ring, as shown in Fig. 7.17. Balancing holes are drilled from the pump inlet through the hub, near the shaft, to allow for leakage past the rear sealing sleeve. The balancing holes must be large enough to establish pump inlet pressure in the rear clearance space near the shaft, in the balance chamber, and to assure that the major pressure drop occurs across the sealing sleeve. If the axial pressure forces

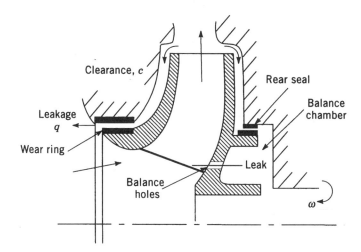

Figure 7.17 Balancing of axial thrust on impeller: leakage flow.

on the impeller are not entirely balanced, the force on the thrust bearing will increase with the impeller exit static pressure and will generally increase with decreasing flow rate.

Some impellers are provided with shallow radial or inclined ridges or vanes on their back side which produce a low pressure in the clearance space near the shaft. Such expeller blades, or pump-out blades, are frequently found on slurry pumps or pumps handling contaminated water. The expeller blades reduce the pressure that acts on the shaft seals. Sometimes, clean water is fed to the other side of the shaft seal to create a leakage flow into the pump and avoid an outward flow of the pumped liquid, which might carry contaminants into the seals. Naturally, expeller blades also change the axial pressure balance of the impeller, which must be taken into account.

LEAKAGE

The flow pattern in the front clearance space of the pump, between the shroud and the housing, is complicated by the fact that on the outer periphery, it connects with the fluid at the impeller exit, and at the inner periphery, with the pump inlet. A sleeve with close clearance, a wear ring at the pump inlet, restricts the leakage flow from the pump exit back to the inlet (Fig. 7.17). This leakage flow is a drain on the high-energy flow delivered by the pump, and therefore reduces the pump efficiency. The leakage flow rate must be deducted from the flow rate through the impeller to arrive at the actual, net flow rate delivered by the pump. Leakage flow is usually kept to 1 or 2 % of the total flow and may or may not be negligible.

Leakage calculations assume that the pressure drop across the wear ring clearance alone governs the leakage flow rate q. A simple orifice flow equation will suffice to estimate the leakage flow rate:

$$q = C_d 2\pi R_w c (2gh_w)^{1/2}$$

In this expression C_d is the discharge coefficient, R_w the wear ring radius, c the clearance width, and h_w the head drop across the wear ring, which can be calculated from the expression derived previously for estimating the axial pressure force on the impeller:

$$h_w = (h_2 - h_1) - \frac{(\tfrac{1}{8})\omega^2 (R_2^2 - R_w^2)}{2g}$$

$$h_2 - h_1 = H_{th} - \frac{C_2^2}{2g} + \frac{C_1^2}{2g}$$

Here $h_2 - h_1$ corresponds to the static head rise across the impeller, calculated using Bernoulli's equation from the theoretical head H_{th} and the velocity heads of the absolute inlet and exit velocities, C_1 and C_2. The impeller exit radius is R_2. The discharge coefficient, C_d, depends on the clearance length, and decreases from 1.0 to 0.5 as the ratio of the clearance length and clearance width increases to about 80 (Bilgen et al. 1973).

Obviously, the smallest possible clearance c is desirable to minimize leakage losses. It is limited by the shaft runout, or eccentricity, which in turn depends on the manufacturing tolerance stackup and on the shaft deflection. Because of its close clearance, the wear ring can act as a water-lubricated bearing and can generate considerable bearing forces, which must be considered in shaft dynamic calculations. Recommended clearance sizes can be found in the standards of the Hydraulic Institute, ASME, ISO, and API.

REFERENCES

Agostinelli, A., Nobles, D., Mockridge, C. R. (1960): An Experimental Investigation of Radial Thrust in Centrifugal Pumps, *ASME Journal of Engineering for Power*, April, pp. 120–126.

Bayly, B. J., Orszag, S. A. (1988): Instability Mechanisms in Shear-Flow Transition, *Annual Review of Fluid Mechanics*, Vol. 20, pp. 359–391.

Busemann, A. (1928): Das Förderverhältniss radialer Kreiselpumpen mit logarithmischspiraligen Schaufeln, *Zeitschrift für Angewandte Mathematik und Mechanik*, Vol. 8, October, pp. 372–384.

Cornell, W. G. (1968): The Stall Performance of Cascades, *Proceedings of the 4th U.S. National Congress on Applied Mechanics*, ASME, New York, pp. 1291–1299.

Csanady, G. T. (1962): Radial Forces in a Pump Impeller Caused by a Volute Casing, *ASME Journal of Engineering for Power*, October, pp. 337–340.

Dallenbach, F. (1961): *Aerodynamic Design and Performance of Centrifugal and Mixed Flow Compressors*, SAE Paper 268 A, January, pp. 9–13.

Dean, R. C., Jr., Senoo, Y. (1960): Rotating Wakes in Vaneless Diffusers, *ASME Journal of Basic Engineering*, September, pp. 563–574.

Due, H. F., Jr. (1966): An Empirical Method for Calculating Radial Pressure Distribution on Rotating Disks, *ASME Journal of Engineering for Power*, April, 188–196.

Foa, J. W. (1958): Pressure Exchange. *ASME Applied Mechanics Review*, December, pp. 655–657.

Foa, J. W. (1973): *Cryptosteady-Flow Energy Separation*, ASME Paper 73-FE-24.

Guelich, J., Florjancic, D., Pace, S. E. (1989): Influence of Flow Between Impeller and Casing on Part-Load Performance of Centrifugal Pumps, in *Pumping Machinery*, ASME FED Vol. 81, pp. 227–235.

Hagelstein, D., Hillewaert, K., Van den Braembussche, R. A., Egenda, A., Keiper, R., Rautenberg, M. (1999): *Experimental and Numerical Investigation of the Flow in a Centrifugal Compressor Volute*, ASME Paper 99-GT-79. Also *Transactions of the ASME*.

Hamkins, C. (1984): *Correlation of One-Dimensional Centrifugal Pump Performance Analysis Method*, ASME Paper 84-WAFM-10.

Hasinger, S. H., Kehrt, I. G. (1963): Investigation of the Shear-Force Pump, *ASME Journal of Engineering for Power*, July, pp. 201–207.

Iversen, H. W., Rolling, R. E., Carlson, J. J. (1960): Volute Pressure Distribution, Radial Force on the Impeller, and Volute Mixing Losses of a Radial Flow Centrifugal Pump, *ASME Journal of Engineering for Power*, April, pp. 136–144.

Johnston, J. P., Rothe, P. H. (1967): Effects of System Rotation on the Performance of Two-Dimensional Diffusers, *ASME Journal of Fluids Engineering*, September, pp. 422–430.

Johnston, P. J., Halleen, R. M., Lezius, D. K. (1972): Effects of Spanwise Rotation on the Structure of Two-Dimensional Fully Developed Channel Flow, *Journal of Fluid Mechanics*, Vol. 56, No. 3, pp. 533–557.

Lorett, J. A., Gopalakrishnan, S. (1986): Interaction Between Impeller and Volute of Pumps at Off-Design, *ASME Journal of Fluids Engineering*, March, pp. 12–18.

McDonald, G. B., Howard, J. H. G. (1973): *The Computation and Utilization of Buseman's Analysis of Potential Flow in an Impeller*, ASME Paper 73-GT-45.

Nece, R. E., Daily, J. W. (1960): Roughness Effects on Frictional Resistance of Enclosed Disks, *ASME Journal of Basic Engineering*, September, pp. 553–562.

Roddy P. J., Darby, R., Morrison, G. L., Jenkins, P. E. (1987): Performance Characteristics of a Multiple-Disk Centrifugal Pump, *ASME Journal of Fluids Engineering*, March, pp. 51–57.

Rohatgi, U., Reshotko, E. (1974): *Analysis of Laminar Flow Between Stationary and Rotating Disks with Inflow*, NASA Report CR-2356, February.

Schlichting, H. (1960): *Boundary Layer Theory*, McGraw-Hill, New York, pp. 86, 179.

Schwarz, D., Wesche, W. (1978): Radial Thrust on Double-Entry Single-Stage Centrifugal Pumps, *Sulzer Technical Review*, 2/1978, pp. 63–68.

Tuzson, J. (1977): Stability of a Curved Free Streamline, *ASME Journal of Fluids Engineering*, September, pp. 603–605.

Tuzson, J. (1993): Interpretation of Impeller Flow Calculations, *ASME Journal of Fluid Engineering*, September, pp. 463–467.

Wiesner, F. J. (1967): A Review of Slip Factors for Centrifugal Impellers, *ASME Journal of Engineering for Power*, October, pp. 558–572.

Wislicenus, G. F. (1965): *Fluid Mechanics of Turbomachinery*, Vol. 2, Dover, New York, p. 626.

Zimmermann, H., Firsching, A., Dibelius, G. H., Ziemann, M. (1986): *Friction Losses and Flow Distribution for Rotating Disks with Shielded and Protruding Bolts*, ASME Paper 86-GT-158.

8
BOUNDARY LAYERS IN PUMPS

BASIC CONCEPT

In fluid machinery, handling water, shear forces are small and localized. Shear forces appear when the total energy of the fluid is different on neighboring streamlines. If the streamlines are straight and parallel, the velocities will be different. A velocity gradient will exist perpendicular to the streamlines. In this case the effect of viscosity manifests itself in a shear stress, force per unit area, which is proportional to the dynamic viscosity μ and to the velocity gradient, the velocity difference over a unit distance perpendicular to the streamlines. A shear stress will also appear if turbulent, unsteady flow fluctuations are present that transfer energy between the streamlines. Such a turbulent shear stress will also be proportional to the velocity gradient. However, the constant of proportionality, taking the place of the dynamic viscosity, will not be a material constant but will depend on the local intensity of turbulent fluctuations and may vary throughout the flow.

Shear forces become important only when high-velocity gradients exist perpendicular to the streamlines. In centrifugal pumps handling water, such locations can be found primarily near the flow passage walls. The fluid in the immediate proximity of the wall does not move. The velocity parallel to the wall increases with distance from the wall and reaches the main flow velocity at a small distance from the wall, as shown in Fig. 8.1. Therefore, fluid shear and the corresponding phenomena are restricted to a small layer near the wall. Generally speaking, these small layers came to be called *boundary layers*.

Since these shear effects are restricted to small regions of the fluid, scientists had the idea that fluid shear and corresponding energy transfer between

92 BOUNDARY LAYERS IN PUMPS

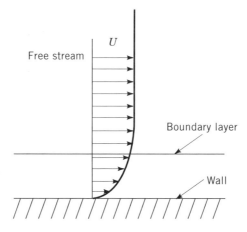

Figure 8.1 Boundary layer.

streamlines needed to be considered only near passage walls, leaving the rest of the fluid unaffected. Since the layer thickness was small and the flow velocity was essentially parallel to the wall, important simplifying assumptions could be made in the fundamental equations of fluid mechanics. Such simplifications in turn led to important analytical solutions. The principal boundary layer assumptions are that the variation of the flow velocity along the wall is negligibly small compared with the velocity variations perpendicular to the wall, and that the static pressure does not vary perpendicular to the wall.

Boundary layers affect the flow in pumps in two respects. Shear forces at the wall produce friction and a loss of energy, which is reflected in the energy conversion efficiency of the pump. Boundary layers also restrict the cross section of the flow passage. Such a blockage should be taken into account when the relationship between flow rate and flow velocity is established. Before the advent of boundary layer calculations, empirical data and relationships were accumulated in the pump industry to account for these effects. Since these guidelines were based on extensive testing, they remain valid and are still widely used.

The general approach of boundary layer calculations consists of starting with an ideal flow calculation which assumes that the velocity at the wall is equal to the velocity at large in the flow. Having an estimate of the velocity at a large distance from the wall, the equations incorporating the assumptions of boundary flow are applied near wall to calculate the shear at the wall and the blockage that the boundary layer produces. A correction can then be applied to the main flow, which takes into account the presence and the effects of the boundary layer (Comolet 1994, Schlichting 1960, White 1991).

TURBULENCE

Turbulence consists of unsteady velocity fluctuations in the flow. Such fluctuations can transfer energy between neighboring streamlines. If the energy on neighboring streamlines is different, if a shear flow exists, turbulent fluctuations can act in a fashion similar to that of the kinematic viscosity of the fluid. Simplified calculations often use an eddy viscosity which replaces the viscosity coefficient.

Under highly viscous flow conditions, flow fluctuations tend to damp out rapidly, and the energy of fluctuations dissipates by viscosity. In the case of low viscosity, turbulent fluctuations persist and are convected far downstream by the flow. The relative importance of viscous forces compared with inertia forces is expressed in the form of a Reynolds number, formed by the product of a typical velocity and a typical length dimension divided by the kinematic viscosity of the fluid. Low Reynolds numbers correspond to viscous flow conditions, whereas at high Reynolds numbers, inertia forces predominate.

Turbulent fluctuations originate with shear layers that become unstable and spill random fluctuations into the flow. Shear layers typically start at a stagnation point on a solid body immersed in the flow or on the boundaries of a closed flow passage. The typical velocity entering into the Reynolds number is the velocity difference between the free stream and the body or wall. Since the velocity at the wall is zero, the velocity entering into the Reynolds number is the relative velocity with respect to the wall. The typical length dimension characterizes the geometry of the flow, the length or width of the object, or the flow passage width. Obviously, the size of the flow oscillations must be limited by the passage walls, for example. The typical length determines the scale of the flow pattern.

The difficulty in calculating the effect of turbulence is that the local intensity of turbulence, and therefore the apparent viscosity or eddy viscosity, depends on the turbulence generated upstream from the location that is being considered. Aircraft flying in quiet air can encounter practically no turbulence. Great difficulties were met in building such wind tunnels, for testing airplane components, which would have sufficiently low levels of turbulence. The upstream turbulence entering a pump, for example, might be quite different depending on whether the pump inlet is submerged in a quiet pool or is preceded by a long pipe. It has been shown experimentally, and also theoretically, that upstream turbulence disturbances are amplified, even in a uniform parallel flow, when the Reynolds number exceeds a critical value, typically about 3000. The usual turbulence model assumes that initially relatively large scale fluctuations are generated, which break up into smaller eddies, and finally dissipate as they are convected downstream. The major portion of the fluctuating energy, which corresponds to the energy loss of the flow, is contained in the large eddies.

Turbulent boundary layer calculations usually assume one of two upstream conditions: Either no turbulence is assumed to be present upstream, or turbulence conditions corresponding to well-established pipe flow are assumed. In a long pipe or uniform flow passage the phenomena of turbulence generation and dissipation reach equilibrium. The distribution of velocities and turbulent fluctuations remains the same at all passage cross sections and are consequently, well defined. The expressions for calculating turbulent boundary layer thicknesses, or blockage, and wall shear assume well-established turbulent pipe flow. They correspond to the generally accepted pipe flow friction loss correlations used throughout industry.

BOUNDARY LAYER THICKNESS

In the simplest case that can be imagined, a flat plate is positioned parallel to a flow of uniform velocity U (Schlichting 1960). For an order-of-magnitude estimate of the boundary layer thickness, the technique of dimensional analysis will be applied, whereby typical values of all the variables affecting boundary layer flow are combined into dimensionless groups, and a relationship between these groups is stipulated on dimensional grounds. In the case of the semi-infinite flat plate, we are missing a typical length dimension. The only reasonable assumption is that the boundary layer thickness depends on the distance from the leading edge of the flat plate L, shown in Fig. 8.2. Since the boundary layer cannot exist ahead of the leading edge where it begins, it is reasonable to assume that the boundary layer thickness δ will gradually grow from the leading edge on. The forces acting in the fluid are the viscous and inertia forces. The corresponding variables are the dynamic viscosity μ, having the dimensions (force × time/length^2), and the density ρ, having the dimensions (force × time^2/length^4). Combining these variables into dimensionless groups it is found that the boundary layer thickness divided by the plate length δ/L must be a function of the Reynolds number which is formed from the free-stream velocity U, the plate length L, and the kinematic viscosity $\nu = \mu/\rho$. The kinematic viscosity of water, at standard ambient conditions, is $\nu = 10.9 \times 10^{-6}$ ft^2/sec.

$$\frac{\delta}{L} \propto \frac{LU}{\nu} = \text{Re}$$

Indeed, analytical solutions of the equations of viscous fluid flow with boundary layer simplifications show that for low Reynolds number, viscous flow conditions, the boundary layer thickness on such a flat plate is given by the relationship

$$\frac{\delta}{L} = 5.0 \left(\frac{\nu}{LU}\right)^{1/2} = 5.0 \,\text{Re}^{-1/2}$$

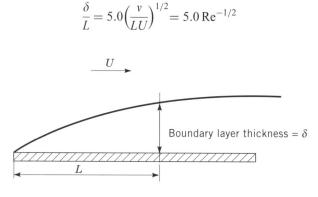

Figure 8.2 Boundary layer on flat plate.

Since the velocity in the boundary layer approaches the free-stream velocity gradually and asymptotically, as the distance from the wall increases, the question arises as to how the boundary layer thickness should be defined. In the expression above, the boundary layer thickness δ corresponds to the distance from the wall where the velocity reaches 99% of the free-stream velocity U.

A more practical definition of the boundary layer thickness is the displacement thickness $= \delta_d$, or blockage, which corresponds to the flow cross-sectional area restriction caused by the boundary layer (Fig. 8.3). For viscous flow it is given by

$$\frac{\delta_d}{L} = 1.72 \, \text{Re}^{-1/2}$$

In the case of high Reynolds number, turbulent flow, where unsteady flow fluctuations transfer energy between streamlines and are responsible for shear, a similar dimensionless relationship applies:

$$\frac{\delta}{L} = 0.37 \left(\frac{LU}{v}\right)^{-1/5} = 0.37 \, \text{Re}^{-1/5}$$

Comparing the two expressions for the viscous and turbulent boundary layer thicknesses, it appears that the transition from viscous flow to turbulent flow occurs at a Reynolds number of $\text{Re} = 3.2 \times 10^5 = 320{,}000$. In reality, the transition from viscous to turbulent flow conditions depends on upstream disturbances in the flow and occurs over a range of Reynolds numbers. Therefore, the numerical value derived above cannot be taken rigorously. In the case of a turbulent boundary layer the displacement thickness becomes even smaller: $\delta_d = \delta/8$. As an example, water flowing under turbulent conditions ($\text{Re} = 1.35 \times 10^7$) at a velocity of 10 ft/sec (3 m/s), would take 15 ft (4.6 m) of

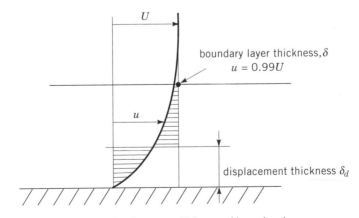

Figure 8.3 Displacement thickness of boundary layer.

length to develop a boundary layer thickness $\delta = 2.5$ in. (6.35 cm), and a displacement thickness $\delta_d = 0.3125$ in. (0.8 cm).

The boundary layer development calculated above would be expected to apply to the blades of impellers and vanes of vaned diffusers. An added complication might arise if the flow velocity approaching the leading edge of the blade is not in a perpendicular plane to the leading edge, as in the case of swept wings, for example. In such a case a velocity component will exist along the leading edge which will convect boundary layer fluid along the leading edge, usually toward the hub of the impeller. Three-dimensional boundary layer calculations have been developed because of the interest in flow predictions for swept airplane wings. However, no closed-form results can be obtained, and only high-speed computers can lead to the solution of specific cases (Humphreys 1981, Tassa et al. 1982).

Boundary layer assumptions break down near the leading edge of blades or vanes if the velocity of the approaching flow is not aligned with the blade. The flow arrives at an angle to the direction of the leading edge, and the flow separates immediately at the leading edge. The flow can be modeled by assuming a separated wake region and a free streamline or shear layer at its boundary. The wake thickness can be calculated assuming an ideal flow pattern and using conformal transformations (Cornell 1962). The subsequent stability and rate of mixing of the shear layer on the wake boundary will depend on the direction of the acceleration perpendicular to the blade surface. Typically, the mixing will be accelerated if separation takes place on the side of the blade facing the direction of rotation—the pressure side—and will be slowed on the trailing, suction side. Consequently, pressure-side leading-edge separation, at flow rates greater than design flow, may reattach, but suction-side separation, at flow rates less than design flow, will tend to persist.

Separation on swept leading edges shows further peculiarities, because a leading-edge vortex may appear in the separated region immediately behind the leading edge, which will trail away along the leading edge, usually toward the hub side of pump impellers. Such a vortex changes entirely the configuration of the separated region.

BOUNDARY LAYERS IN CLOSED FLOW PASSAGES

The flat plate boundary layer calculation also applies to closed passages, as for example the inlet of a pipe, as long as the pipe diameter or distance to the opposite wall of the flow passage is large compared with the thickness of the boundary layer. The boundary layer "does not know" that there is a wall at a relatively great distance. When the pipe diameter or passage width becomes comparable with the length L along the wall, it starts to influence the boundary layer development. In the dimensional analysis approach the passage width becomes the characteristic length dimension, and the Reynolds number based on the passage diameter or width becomes dominant.

BOUNDARY LAYERS IN CLOSED FLOW PASSAGES

In the case of a flat plate the fluid friction energy dissipated at the wall is taken at the expense of the kinetic energy of the fluid in the boundary layer. The layer has to grow, and more and more slow-moving fluid accumulates in the boundary layer. In a closed passage a steady condition must be reached, since the boundary layers cannot grow indefinitely. Therefore, in a closed passage the wall friction energy loss is compensated by a pressure loss along the passage. When the boundary layer thickness becomes significant compared with the passage width, the velocity profile across the passage gradually converges to a steady configuration. At low Reynolds numbers, under viscous flow conditions, the fully developed velocity profile will be parabolic (Fig. 8.4). Under turbulent conditions the velocity in the central portion of the flow passage will be almost uniform, and near the walls the velocity will drop off over a small distance, a steady boundary layer.

In a circular pipe, at high Reynolds numbers, the steady, local flow velocity u becomes a function of the distance from the wall y toward the center of the pipe at a radius R where the centerline velocity is U:

$$\frac{u}{U} = \left(\frac{y}{R}\right)^{1/n}$$

The exponent n depends on the Reynolds number and takes on the value of $n = 6$ at $Re = 4 \times 10^3$, $n = 7$ at $Re = 10^4$, and reaches $n = 10$ at $Re = 3 \times 10^6$. The

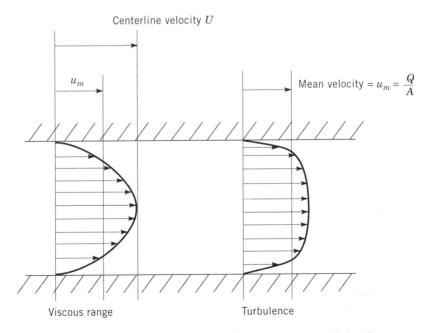

Figure 8.4 Velocity profiles in a circular pipe in the viscous and turbulent flow range.

ratio of the mean velocity u_m, which would correspond to the flow rate divided by the pipe cross section and the maximum centerline velocity U, becomes

$$\frac{u_m}{U} = \frac{2n^2}{(n+1)(2n+1)}$$

It will be noted that this ratio increases with the Reynolds number from 0.791 at $n = 6$ to 0.865 at $n = 10$. The difference between the calculated value and 1 corresponds to the fraction of the flow passage cross section that is taken up by boundary layer blockage.

SKIN FRICTION

From the point of view of pumping machinery analysis and design, the significance of boundary layer calculations is that they allow a theoretically based estimate of wall friction losses in flow passages. By definition, the viscous shear stress τ, shear force per unit area, at the flow passage wall is proportional to the velocity gradient at the wall. In the case of a flat plate, an exact analytical solution of the viscous fluid flow equations, with boundary layer simplifications, led to the following expression of the local shear stress, where the Reynolds number is based on the length from the leading edge of the flat plate:

$$\frac{\tau}{\rho U^2/2} = 0.664\, \mathrm{Re}^{-1/2}$$

In the case of turbulent flow in a pipe, the local shear stress at the wall becomes

$$\frac{\tau}{\rho U^2/2} = 0.0445\, \mathrm{Re}^{-1/4}$$

In this expression the Reynolds number is based on the pipe radius. The $-\frac{1}{4}$ exponent was derived for the case when the turbulent velocity distribution in the pipe correlates with the seventh power, $n = 7$, of the Reynolds number. Extrapolating to velocity distributions at higher Reynolds numbers, the exponent in the expression of the wall shear would become $-2/(n+1)$. These shear stress values correspond to the well-known pipe friction coefficients.

BOUNDARY LAYERS ON ROTATING DISKS

The sides of a rotating radial impeller can be considered rotating disks, provided that the width of the space between the impeller faces and the stationary housing is much larger than the boundary layer. Consequently, this case has a great importance for pump design and operation. Figure 8.5 illustrates the tangential

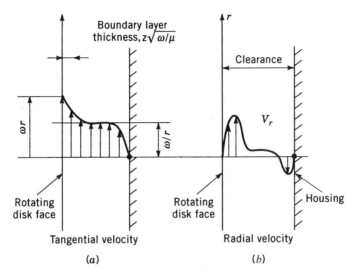

Figure 8.5 (a) Tangential and (b) radial velocity profiles in the clearance space between rotating impeller face and stationary housing.

and radial velocity distribution in the clearance space between the rotating impeller face and the housing wall.

It so happens that an exact solution of the complete fluid mechanic equations, the Navier–Stokes equations, can be obtained for a disk rotating in an infinite fluid if the Reynolds number based on the disk radius, the rotational speed, and the dynamic viscosity of water remains small, less than 50,000 (Schlichting 1960, p. 83). The angular velocity ω of the disk and the kinematic viscosity v govern the thickness of the boundary layer, which must have the form $(\omega/v)^{1/2}$ to balance the dimensions. The boundary layer thickness corresponds to the distance from the disk where the tangential velocity becomes half of the disk velocity. In this particular case the tangential velocity at great distance from the disk is assumed to be zero. At the same distance from the disk the radial velocity induced by the rotation of the disk reaches a maximum and becomes $u = 0.180 r\omega$. The total radial flow generated by the disk of radius R is $Q = 0.443\pi R^2 (v\omega)^{1/2}$. For these laminar conditions the disk friction torque from one side of the disk becomes

$$C_M \left(\frac{\rho \omega^2 R^5}{2} \right) = \frac{1.935}{\text{Re}^{-1/2}} \frac{\rho \omega^2 R^5}{2}$$

$$\text{Re} = \frac{\omega R^2}{v}$$

The coefficient C_M stands for the disk friction torque in dimensionless form. Its numerical value does not depend on the type of measurement system used, English or SI.

For turbulent conditions boundary layer calculations become more complex. The boundary layer thickness δ, the flow rate Q, and the torque coefficient C_M are found to be

$$\delta = 0.526 r^{3/5} \left(\frac{\nu}{\omega}\right)^{1/5}$$
$$Q = 0.109 \omega R^3 \cdot \text{Re}^{-1/5}$$
$$C_M = 0.073 \, \text{Re}^{-1/5}$$

The expressions above are for a disk in an infinite fluid. If the disk is enclosed in a housing with a side clearance s, which is smaller than the boundary layer, the moment for one side of the disk becomes for laminar conditions:

$$C_M = \frac{\pi(R/s)}{\text{Re}}$$

If the clearance is a multiple of the boundary layer, the moment coefficients for laminar and turbulent flow become

$$C_M = \begin{cases} \dfrac{1.33}{\text{Re}^{1/2}} \\ \dfrac{0.0311}{\text{Re}^{1/5}} \end{cases}$$

Several detailed studies have been devoted to the case when the rotating disk is enclosed in a housing (Bakke et al. 1973, Bodonyi and Stewartson 1975, Chanaud 1971, Cooper 1973, Maroti et al. 1960, Rohatgi and Reshotko 1974, Schilling et al. 1974, Senoo and Hayami 1976). In this case the tangential velocity, gradually decreasing perpendicular to the disk, approaches a value that is about half of the circumferential speed of the disk midway between the disk and the housing wall. A boundary layer also exists at the wall, as described in connection with the disk friction loss calculation in Chapter 7. The analytical expressions above remain valid except for the numerical coefficients, which may take on other values, depending on the geometrical details.

Very detailed experimental torque measurements have been made on enclosed rotating disks which closely conform to the theoretical results. These data are being used in pump design and performance prediction to estimate disk friction torque on impellers.

BOUNDARY LAYERS IN VANELESS DIFFUSERS

Centrifugal pumps often use vaneless diffusers at the exit of the impeller if sufficient radial space is available. They consist of parallel radial walls, and unlike vaned diffusers, are very tolerant to variation in flow rates, since in the absence of vanes, no incidence losses can occur. The flow from the impeller exit enters the vaneless diffuser with a certain tangential and radial velocity which slows down toward the diffuser exit at a greater radius.

A loss-free fluid flow analysis predicts velocities based on three fundamental equations. The radial variation of the tangential velocity is governed by the principle of conservation of angular momentum. Consequently, the tangential velocity is expected to decrease in inverse proportion to the radius:

$$V_t = \frac{\text{angular momentum}}{r}$$

The radial velocity must satisfy the continuity equation and maintain constant flow rate through consecutive flow cross sections, the area being given by the product of the diffuser width s and the circumference at any radius: $V_r = Q/2\pi rs$. Consequently, the radial velocity also would be expected to decrease in proportion to the radius.

The pressure can be obtained from the equation of radial force balance, which states that the incremental radial pressure increase will be equal to the centrifugal force, the fluid density multiplied by the square of the tangential velocity and divided by the radius:

$$\frac{dp}{dr} = \frac{\rho V_t^2}{r}$$

Since the equation of conservation of energy, Bernoulli's equation, is an alternative to the basic equations, the pressure can also be calculated from the changes of the tangential and radial velocity heads. In loss-free flow the increase in pressure head must correspond to the decrease in tangential and radial velocity heads. This ideal flow pattern serve to define the free-stream conditions of a boundary layer calculation (Gopalakrishnan 1993, Jansen 1964, Stanitz 1954).

When boundary layers and viscous or turbulent shear are present, as shown in Fig. 8.6, the torque from the tangential wall shear forces will oppose the conservation of angular momentum. If well-established turbulence levels are assumed, the tangential velocity distribution on either side of the vaneless diffuser will follow the configuration measured in pipe flow. Such a turbulence level assumption might be questioned, because the flow leaving the impeller is highly turbulent. Depending on the blade loading at the impeller blade tips, the relative velocity on the pressure side of the blades is lower than on the suction side, which results in a periodic absolute velocity fluctuation at the impeller exit. Whatever the turbulence level and the corresponding shear stresses might be, qualitatively speaking, the maximum centerline tangential velocity in the vaneless diffuser, as well as the average velocity and angular momentum, will gradually decrease with the radius.

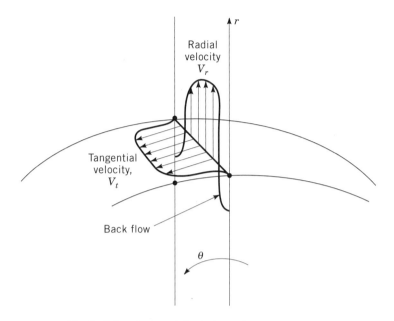

Figure 8.6 Radial and tangential velocity profiles in a vaneless diffuser.

A different situation is met in the radial direction. The average radial velocity must maintain a constant flow rate despite increasing boundary layer blockage. In addition, the radial pressure increase on the centerline, where high tangential velocities prevail, will be greater than near the walls, where the tangential velocity approaches zero. Such a difference in pressure rise must be compensated by secondary flows radially inward near the walls, opposing boundary layer flow. Consequently, the radial boundary layer will thicken considerably, and the point will be reached when the boundary layers from the two walls merge at the center of the flow passage. No or little pressure recovery can be expected beyond this point.

Calculations (Gopalakrishnan 1993) supported by experimental data (Jansen 1964) show that depending on the impeller exit flow angle, or the ratio of radial to tangential velocities at the diffuser inlet, vaneless diffusers with radius ratios, ratios of exit to inlet radii beyond about 1.5 to 2.0 do not recover additional pressure head and will not improve the efficiency of a centrifugal pump.

BOUNDARY LAYERS INSIDE ROTATING IMPELLERS

Boundary layer development inside impellers resembles the conditions in a vaneless diffuser provided that calculations are based on the relative tangential and radial velocities measured with respect to the walls of the impeller (Fig. 8.7). However, there is a major, important difference between the boundary layers on

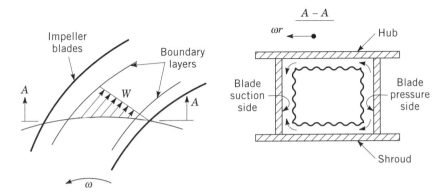

Figure 8.7 Boundary layers in the rotating impeller; secondary flow in the boundary layers due to Coriolis acceleration.

stationary walls in a swirling flow and those in the impeller. Because of the Coriolis acceleration, a circumferential pressure gradient exists. The pressure decreases in the tangential direction from the side of the blades facing in the direction of rotation toward the back side of the next blade. The pressure rise is proportional to the Coriolis acceleration, which is equal to the product of the angular velocity ω and the radial velocity W_r. Since the relative velocity vanishes in the boundary layers on the hub and shroud flow passage walls—unlike in the main flow—little or no tangential pressure rise exists along these walls. This lack of pressure rise must be compensated by secondary currents in the boundary layers from the pressure side of the blade toward the suction side. These currents carry boundary layer fluid from the pressure side of the blades to the suction side, where they accumulate and thicken the boundary layer. Because of these secondary currents, arising from the action of the Coriolis acceleration, which will exist only in rotating impellers but will not exist when the impellers do not rotate, boundary layer thickening and separation will not appear on the pressure side of impeller blades but on the suction side.

In addition to the effects noted above, the Coriolis acceleration also affects the distribution of the turbulence intensity. It can amplify or dampen turbulent fluctuations, depending on the local flow conditions. Therefore, the usual, fully developed pipe flow turbulence assumptions cannot be accepted with confidence, and special corrections are needed (Howard et al. 1980). Similar transverse acceleration effect, such as that resulting from streamline curvature, can strongly influence boundary layer development (Ellis and Joubert 1974, Irwin and Arnot Smith 1975, Johnston 1972, Litvai 1972, So and Mellor 1973, Wollkind and DiPrima 1973).

It is important to note that to a first order of approximation, the flow in rotating impellers can be calculated as if the impeller were stationary and would not rotate. However, if real viscous flow conditions are to be taken into account, impeller rotation and the action of the Coriolis acceleration must be introduced into the analysis. Ideal loss-free flow calculations sometimes show that the flow would separate on the pressure side of the blades. These calculations give a false picture of the flow pattern. In reality, the flow tends to separate on the suction side of the blades.

Boundary layers in rotating impellers present another peculiarity. In stationary flow passages the fluid in the boundary layer has lost some, if not all, of its energy. It is customary to refer to boundary layer fluid as stagnant, low-energy fluid. In rotating impellers, boundary layer fluid does not necessarily have less energy than the main stream. From the point of view of a stationary observer a fluid element in the impeller has an energy proportional to its angular momentum, its absolute tangential velocity multiplied by the radius from the center of rotation. In impellers with backward-curved blades, the absolute tangential velocity of the fluid is generally less than the circumferential velocity of the impeller itself. Indeed, the fluid in the impeller flows in the direction opposite the sense of rotation with respect to the impeller. However, boundary layer fluid which hardly moves with respect to the impeller, and practically rotates at the same speed as the impeller, has a higher absolute tangential velocity than that of the main stream. Consequently, boundary layer fluid in impellers with backward-curved blades has more energy than the main stream. Therefore, the boundary layer fluid leaving the impeller can add to, instead of subtracting from, the theoretical head, the energy that the impeller imparts to the flow.

Boundary layer flow from the blade pressure and suction faces will mix with the main flow at the impeller exit. The boundary layers on the inner hub and shroud faces of the impeller will produce trailing vortices at the impeller exit. Indeed, since the energy imparted to the fluid by the impeller, the lift produced by the blades, will be different in the boundary layers than in the main stream, streamwise vorticity will be spilled into the flow, which will roll up into trailing vortices in the corners where the blades meet the internal hub and shroud faces of the impeller. The trailing vortices become particularly evident in slurry pumps, in which the particles carried by the vortices can wear deep grooves into the impeller.

PRACTICAL PUMP DESIGN CALCULATIONS

Because of the great diversity and complexity of turbulent boundary layer flow in pumps, very detailed, lengthy, and expensive calculations of uncertain validity are often not warranted. Empirical data from pumps, similar to those being considered, become extremely valuable. Dimensionless loss and blockage coefficients can be derived from such test data. Great variations in geometrical scale or viscosity will be reflected in the change of the respective Reynolds number, which enters into the relationship connecting the dimensionless boundary layer thickness or shear stress to the Reynolds number. If a valid qualitative model of the flow pattern can be constructed, the quantitative results, the numerical values, can be calculated from dimensionless correlations with empirical coefficients. Such experimentally derived dimensionless coefficients have been accumulated over years by well-established pump manufacturers. They represent the confidential technological capital of successful companies.

REFERENCES

Bakke, E., Kreider, J. F., Kreith, F. (1973): Turbulent Source Flow Between Parallel Stationary and Co-rotating Disks, *ASME Journal of Fluid Mechanics*, Vol. 58, Pt. 2, pp. 209–231.

Bodonyi, R. J., Stewartson, K. (1975): Boundary-Layer Similarity Near the Edge of a Rotating Disk, *ASME Journal of Applied Mechanics*, September, p. 584. Also ASME Paper 75-WA/APM-3.

Chanaud, R. C. (1971): Measurements of Mean Flow Velocity Beyond a Rotating Disk, *ASME Journal of Basic Engineering*, June, pp. 199–204.

Comolet, R. (1994): *Mécanique expérimentale des fluides*, Tome II: *Dynamique des fluides réels, Turbomachines*, Masson, Paris, pp. 177–227.

Cooper, P. (1973): *Turbulent Fluid Friction of Rotating Disks*, NASA Report CR-2274, July.

Cornell, W. G. (1962): The Stall Performance of Cascades, *Proceedings of the 4th U.S. National Congress on Applied Mechanics*, ASME, New York, pp. 1291–1299.

Ellis, L. B., Joubert, P. N. (1974): Turbulent Shear Flow in a Curved Duct, *ASME Journal of Fluid Mechanics*, Vol. 62, Pt. 1, pp. 65–84.

Gopalakrishnan, S. (1993): Boundary Layer Growth in Vaneless Diffusers, in *Pumping Machinery*, ASME FED Vol. 154, pp. 207–218.

Howard, J. H. G., Patankar, S. V., Bordynuik, R. M. (1980): *Flow Prediction in Rotating Ducts Using Coriolis-Modified Turbulence Models*, ASME Paper 80-WA/FE-1.

Humphreys, D. A. (1981): Three-Dimensional Wing Boundary Layer Calculated with Eight Different Methods, *AIAA Journal*, Vol. 19, No. 2, pp. 232–234.

Irwin, H. P. A. H., Arnot Smith, P. (1975): Prediction of the Effect of Streamline Curvature on Turbulence, *Physics of Fluids*, Vol. 18, No. 6, pp. 624–630.

Jansen, W. (1964): Steady Flow in a Radial Vaneless Diffuser, *ASME Journal of Basic Engineering*, September, pp. 607–619.

Johnston, J. P. (1972): *The Suppression of Shear Layer Turbulence in Rotating Systems*, ASME Paper 72-FE-B.

Litvai, E. (1972): Prediction of Velocity Profiles for Turbulent Boundary Layers on the Blading of Radial Impellers, *Proceedings of the Conference on the Fluid Mechanics of Fluid Machinery*, Budapest, Hungary, pp. 771–782.

Maroti, L. A., Deak, G., Kreith, F. (1960): Flow Phenomena of Partially Enclosed Rotating Disks, *ASME Journal of Basic Engineering*, September, pp. 539–552.

Rohatgi, U., Reshotko, E. (1974): *Analysis of Laminar Flow Between Stationary and Rotating Disks with Inflow*, NASA Report CR-2356, February.

Schilling, R., Siegle, H., Stoffel, B. (1974): Strömung und Verluste in drei wichtigen Elementen radialer Kreiselpumpen, *Strömungsmechanik und Strömunmgsmaschinen*, No. 16, July, pp. 1–46.

Schlichting, H. (1960): *Boundary Layer Theory*, McGraw-Hill, New York, 1960.

Senoo, Y., Hayami, H. (1976): An Analysis on the Flow in a Casing Induced by a Rotating Disk Using a Four-Layer Flow Model, *ASME Journal of Fluids Engineering*, June, pp. 192–198.

So, R. M. C., Mellor, G. L. (1973): Experiment on Convex Curvature Effects in Turbulent Boundary Layers, *ASME Journal of Fluid Mechanics*, Vol. 60, Pt. 1, pp. 43–62.

Stanitz, J. D. (1954): *One-Dimensional Compressible Flow in Vaneless Diffusers*, NACA-TN 2610.

Tassa, A., Atta, E. H., Lemmerman, L. A. (1982): *A New Three-Dimensional Boundary Layer Calculation Method*, AIAA-82-0224.

White, F. M. (1991): *Viscous Fluid Flow*, McGraw-Hill, New York.

Wollkind, R., DiPrima, R. C. (1973): Effect of a Coriolis Force on the Stability of Plane Poiseuille Flow, *Physics of Fluids*, Vol. 16, No. 12, pp. 2045–2057.

9

CAVITATION IN PUMPS

DESCRIPTION

Cavitation has been defined as the phenomenon when the local absolute static pressure of the fluid, somewhere in the flow, drops below the vapor pressure of the fluid and vapor bubbles appear (Arndt 1981; Brennen 1994; *La Houille Blanche* 1988, 1992; Avellan et al. 1995). As the bubbles are convected downstream, they reach a location where the pressure increases again above the vapor pressure. They collapse and create a sudden, very high pressure pulse which can damage the material of a nearby flow passage wall. In centrifugal pumps the lowest pressure usually appears on the impeller blades at the inlet, especially at flow rates higher than the design flow rate, when the flow separates on the leading, pressure side of the blades.

Pump designers wish to know the conditions for the appearance of cavitation in pumps and the extent of the resulting damage. Prediction of these items remains difficult because of the complexity of the phenomena and because several factors affecting cavitation in a certain application remain unknown, such as the air content of the water handled by the pump. Commercially available pumps must be able to accommodate such uncertainties and operate satisfactorily over a range of application conditions. Exact numerical calculations, executed for nominal operating conditions, have uncertain general validity, and an uncertainty margin must be allowed. Therefore, only approximate methods of calculation, supported by empirical data, can be offered here.

In particular, two phenomena concern pump designers and users. Cavitation damages the impeller and cannot be tolerated for long. Extensive cavitation and significant amounts of vapor will impair operation of the pump. This second issue is discussed in detail in Chapter 15.

Pump designers and users have been concerned with cavitation primarily when pumping cold water. Most experimental data and guidelines were taken with and apply to water. However, water has very unusual physical properties affecting cavitation. It has considerably higher latent heat of evaporation and surface tension than those of other liquids. Indeed, it has been noted that cavitation of other liquids is different and less damaging than cavitation of water (Salemann 1959). Cavitation with water is not typical. Since the formation of vapor bubbles involves thermodynamic phenomena, mass and heat transfer to the bubble, several physical properties of the liquid other than vapor pressure affect the cavitation process. It can be assumed that the onset of cavitation, the first appearance of vapor bubbles, indeed depends primarily on the vapor pressure of the liquid. The phenomena of further evolution and collapse of the bubbles, of cavitation damage, and of pump performance with a significant amount of vapor involve other fluid properties and differ for different kinds of fluids. The issues presented in this chapter are restricted to those dealing with cavitation of cold water.

Unfortunately, the inception of cavitation, the first appearance of vapor bubbles, only produces some ill-defined noise which cannot provide a distinct indication for field applications. Practical considerations demanded that the appearance of cavitation be defined by distinct deterioration of the pump performance, which corresponds to an advanced state of vapor bubble formation. The onset of cavitation in centrifugal pumps is determined during tests, while gradually lowering the absolute inlet pressure of the pump. The generally accepted condition, which signals the appearance of cavitation, is when the pump head decreases by 3% compared with its value at high inlet pressures. However, cavitation bubbles appear much sooner with decreasing inlet pressure, and serious damage can also be done. The appearance of the bubbles coincides with the appearance of cavitation noise, a crackling sound, which sounds somewhat like the noise when a fluid just about starts to boil. Attempts have been made recently to empirically correlate the intensity of this noise with cavitation damage. (Guelich 1989 a,b). Only correlations based on the fluid mechanics of the flow are presented here.

NET POSITIVE SUCTION HEAD

A centrifugal pump requires a certain total head h_{01} at the impeller inlet in order to avoid the absolute static pressure at the inlet or on the impeller blades near the inlet dipping below the vapor pressure of the fluid, causing cavitation to appear. The vapor pressure of the fluid, p_{vp}, is a material property which depends on the temperature. When the absolute pressure in the fluid, the pressure measured above absolute vacuum, falls below the vapor pressure, vapor bubbles appear and the fluid starts to boil. Absolute vacuum corresponds to a pressure of 14.7 psi, or a water column pressure of 33.8 ft (10.3 m) below standard ambient pressure. Therefore, pump inlet pressure calculations must be expressed in absolute pressure, psia, measured from absolute vacuum. Ambient pressure then becomes 14.7 psia in terms of absolute pressure.

The vapor pressure of water depends on the temperature; at 70°F it is 0.3629 psia (2500 N/m²) and at 85°F it is 0.5958 psia (4100 N/m²), only approximately 1 ft above absolute vacuum, which corresponds to 0 psia. Organic fluids can have very different vapor pressures (Salemann 1959). If the static wall pressure p_1 is measured at the pump inlet with a gage that measures the pressure in psig relative to ambient conditions, the atmospheric pressure, p_{atm}—about 14.7 psia—must be added to arrive at the absolute value of the static pressure p_1 in psia. The relative static pressure p_1 (psig) can be negative—a suction with respect to ambient pressure—but the absolute pressure will always be positive.

$$p_1(\text{psia}) = p_1(\text{psig}) + p_{atm}(\text{psia})$$

To estimate the pressure needed by the pump at the inlet, the total, absolute pressure p_{01} psia must be calculated, which amounts to the sum of the static pressure p_1 (psia) and dynamic pressure of the absolute inlet velocity C_1:

$$p_{01}(\text{psia}) = p_1(\text{psia}) + \frac{\rho C_1^2}{2}$$

The capability of a pump to avoid cavitation is expressed by the net positive suction head required (NPSHR) by the pump, which is established by pump manufacturers by tests and is usually shown in pump catalogs. Standard testing procedures (ASME. ISO, API, Hydraulic Institute) prescribe that the pressure at the inlet flange of the pump be lowered until the head produced by the pump declines by 3%. The NPSHR value is then based on the absolute total pressure p_{01} psia at the pump inlet minus the absolute vapor pressure of the fluid p_{vp}. The definition of the net positive suction head required (NPSHR) by the pump becomes

$$\text{NPSHR} = \frac{p_{01} - p_{vp}}{\rho g} = \frac{p_1}{\rho g} + \frac{C_1^2}{2g} - \frac{p_{vp}}{\rho g}$$

In this expression p_{01} will be that particular value of the inlet total pressure for which cavitation appears somewhere in the pump. The appearance of cavitation will depend on the flow rate, and so will the particular value of the the total inlet pressure that enters into the expression of NPSHR. It will be observed that cavitation will take place at the inlet flange if the NPSHR of the pump equals the inlet velocity head $C_1^2/2g$. Then the static pressure head at the pump inlet flange $p_1/\rho g$ will be equal to the vapor pressure head of the fluid $p_{vp}/\rho g$.

Flow velocities even higher than at the pump inlet flange can be reached on the blades of the rotating impellers. The local velocity head can be higher and the absolute static pressure can be lower than at the inlet flange. The impeller blades at the inlet of pumps are usually designed to admit the flow into the impeller

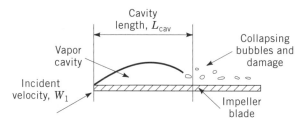

Figure 9.1 Cavitation bubble.

without any change in velocity. Therefore, at design conditions the pressure on the blade leading edge will be essentially the same as at the inlet flange. However, at flow rates different from the design flow rate, especially at higher flow rates, the incoming flow will not be aligned with the direction of the blades, the flow will separate at the blade tips as shown in Fig. 9.1, and a separated flow region will restrict the flow passage cross section between the blades. The flow velocity will have to accelerate to maintain the flow rate. Higher velocities and lower pressures would be expected on the blade leading edge. Indeed, experiments show that the inlet head needed to avoid cavitation, the NPSHR of the pump, increases with increasing flow rate. Pump performance plots in manufacturers' catalogs usually show the experimentally determined NPSHR of the pump as a function of the flow rate. The experimentally measured NPSHR is that value for which the pump head drops 3% compared with the pump head at the same flow rate but with ample inlet head. Experiments are performed at standard ambient temperatures and pressures. However, cavitation bubbles and damage can appear at inlet heads higher than those corresponding to the 3% head drop.

A dimensionless cavitation coefficient σ can be defined as the ratio of the net positive suction head required by the pump, the total inlet head less the vapor pressure head of the fluid, divided by the velocity head of the circumferential velocity at the pump inlet U_1:

$$\sigma = \frac{\text{NPSHR}}{(U_1^2/2g)}$$

This cavitation number has been used in the past by pump designers to obtain some estimate of the likelihood of cavitation. Empirical data have been published which correlate specific speed with the cavitation coefficient (Stepanoff 1957, Sulzer 1989). It applied to the scaling of similar pumps, in particular reflecting the effect of rotational speed on NPSHR. Analytical models are presented below for estimating the conditions at the onset of cavitation in the pump and for arriving at an estimate of the NPSHR required by the pump.

CAVITATION AT DESIGN CONDITIONS

At design conditions the absolute velocity on the blades at the inlet is the same as the velocity at the pump inlet flange C_1. This assumes that an allowance was made for the blockage of the blade thickness when the inlet cross-sectional area was calculated. Blade blockage is neglected below. At the onset of cavitation the absolute inlet static pressure becomes equal to the vapor pressure of the fluid; the NPSHR equals the inlet velocity head:

$$\text{NPSHR} = \frac{C_1^2}{2g}$$

Using the relationships

$$U_1 = \frac{\omega D_1}{2} = \frac{N 2\pi}{60} \frac{D_1}{2}$$

$$C_1 = \frac{Q}{(\pi D_1^2/4)}$$

$$\frac{U_1}{C_1} = \tan \beta_1$$

the impeller inlet diameter can be expressed as

$$D_1 = \left[\left(\frac{8}{\pi} \tan \beta_1 \frac{Q}{N} \frac{60}{2\pi} \right) \right]^{1/3}$$

Consequently, if the consistent units N (rpm) and Q (ft^3/sec) are used, the numerical coefficient becomes a dimensionless pure number, and the expression for the NPSHR in feet takes the form

$$\text{NPSHR} = \frac{C_1^2}{2g} = \frac{(\tan \beta_1)^{-2} U_1^2}{2g} = 0.023 \left(\frac{1}{2g} \right) (\tan \beta_1)^{-4/3} N^{4/3} Q^{2/3}$$

Since the velocities can differ from the average velocity along the leading edge, the numerical coefficient may differ from the value above. Cavitation may occur first near the shroud, for example, where higher velocities may exist. However, an approximate estimate can be obtained simply by assuming an average flow velocity (Vlaming 1989).

In analogy to the specific speed, a suction specific speed S is used in industrial practice to provide a measure of the suction capability of a pump. Retaining consistent units, the suction specific speed also remains dimensionless.

$$S = \frac{N Q^{1/2}}{\text{NPSHR}^{3/4}}$$

112 CAVITATION IN PUMPS

By comparing with the expression of NPSHR above, the suction specific speed can be identified with

$$S = \tan \beta_1 \left(\frac{2g}{0.023}\right)^{3/4}$$

The theoretical derivation of the NPSHR above corresponds to the first inception of cavitation. The suction specific speed values in catalogs derive from experiments and correspond to the 3% head drop value and include the flow rate in gpm. Therefore, they can differ considerably from the calculated values.

In general industrial practice, the NPSHR is sometimes expressed in terms of the specific speed N_s, in the form

$$\text{NPSHR} = C_c \frac{1}{2g} N^{4/3} Q^{2/3} = C_c \frac{1}{2g} N_s^{4/3} H$$

$$N_s = \frac{NQ^{1/2}}{H^{3/4}}$$

The expression must have this particular form to balance the dimensions, as can easily be verified. The coefficient C_c depends on the inlet blade angle β_1. It can be dimensionless if the dimensionless specific speed is used and the speed, flow rate, and head are expressed in consistent units. It will not be dimensionless if the specific speed is expressed in the usual English units.

The Hydraulic Institute published plots showing the approximate NPSHR required for pumps. A satisfactory correlation is found for the empirical data when the exponent of the specific speed is increased to $\frac{5}{3} = 1.666$. The usual inlet blade angle of commercial pumps tends to change with specific speed and introduces a change in the coefficient of proportionality by the factor $N_s^{1/3}$. The correlation to match the empirical data becomes

$$\text{NPSHR} = 0.415 \times 10^{-6} N_s^{5/3} H$$
$$= 0.415 \times 10^{-6} N_s^{1/3} N^{4/3} Q^{2/3}$$

Here the specific speed N_s is in English units, Q (gpm), H (ft), and N (rpm). The numerical constant is not dimensionless and will be different if SI units are used. It should be noted that NPSHR does not really depend on the pump head H, which appears in this expression only to compensate for the dependence of the specific speed on the head. When the NPSHR is expressed as a function of the shaft speed N and the flow rate Q, the head H disappears.

The correlation offered by the Hydraulic Institute is based on average inlet designs generally encountered in the industry and gives adequate estimates for the application of pumps, but a more detailed calculation of the inlet velocities will be required for pump design. The correlation, and the data of the Hydraulic Institute, are based on a pump head drop of 3%, water at a temperature of 85°F, and standard ambient pressure. For pump applications the NPSHR estimate must be conservative, because cavitation damage appears long before the head starts to drop off noticeably. Cavitation and NPSH requirements have been investigated experimentally operating on hot water and liquids other than water (Salemann 1959).

CAVITATION AT OFF-DESIGN CONDITIONS

At off-design conditions the inlet velocity is not aligned with the blades, and the flow separates at the blade leading edge, forming a separated wake region at a constant pressure p_s, as discussed in Chapter 7 and in connection with incidence losses included in performance calculations where the flow conditions are illustrated. The flow adjusts gradually, and the relative flow velocity W_s becomes uniform and parallel to the blade surfaces. Since no losses occur up to this location, the sum of the static and velocity heads must still equal the total head and must satisfy the equation of conservation of energy. If W_{sm} and W_{st} are the meridional and tangential components of the relative separated flow velocity W_s, W_{sm} is also the absolute meridional velocity, and $U_1 - W_{st}$ is the absolute tangential velocity. The energy equation, Bernoulli's equation, can be written in terms of the absolute velocity and static pressure at the inlet C_1 and p_1, and the same variables corresponding to the separated flow:

$$\frac{p_{01}}{\rho g} = \frac{p_1}{\rho g} + \frac{C_1^2}{2g} = \frac{p_s}{\rho g} + \frac{W_{sm}^2}{2g} + \frac{(U_1 - W_{st})^2}{2g}$$

The NPSHR value required by the pump will be obtained when the pressure in the separated region p_s becomes equal to the vapor pressure of the fluid, p_{vp}. The NPSHR value equals the total inlet head, less the vapor pressure head, which corresponds to this condition:

$$\begin{aligned} \text{NPSHR} &= \frac{p_{01}}{\rho g} - \frac{p_{vp}}{\rho g} = \frac{p_1}{\rho g} + \frac{C_1^2}{2g} - \frac{p_{vp}}{\rho g} \\ &= \frac{W_{sm}^2}{2g} + \frac{U_1^2}{2g} - \frac{2U_1 W_{st}}{2g} + \frac{W_{st}^2}{2g} \\ &= + \frac{W_s^2}{2g} + \frac{U_1^2}{2g} - \frac{U_1 W_{st}}{g} \end{aligned}$$

Since the relative velocity is parallel to the blade which makes an angle β_1 with the meridional direction, $W_{st} = W_s \sin \beta_1$, the NPSHR becomes

$$\text{NPSHR} = + \frac{W_s^2}{2g} + \frac{U_1^2}{2g} - \frac{U_1 W_s}{g} \sin \beta_1$$

CAVITATION IN PUMPS

The relationship between the inlet relative velocity W_1 and the velocity at separation W_s depends on the size of the separated region and the flow passage blockage it creates. The ratio of these two velocities can be obtained from an expression that was derived for ideal, loss-free separated flow on the blade leading edge for a blade angle β_1 and a flow angle β_{F1} (Cornell 1962):

$$\lambda = \frac{W_1}{W_S} = \frac{\cos \beta_{F1}}{\cos(2\beta_{F1} - \beta_1)} - \left\{ \frac{(\cos \beta_{F1})^2 - (\cos \beta_1)[\cos(2\beta_{F1} - \beta_1)]}{[\cos(2\beta_{F1} - \beta_1)]^2} \right\}^{1/2}$$

The NPSHR value is then given in terms of the inlet relative velocity W_1 by

$$\text{NPSHR} = \frac{1}{\lambda^2}\frac{W_1^2}{2g} + \frac{U_1^2}{2g}\left(1 - 2\sin\beta_1 \frac{W_1}{U_1}\frac{1}{\lambda}\right)$$

The NPSHR value can also be expressed in terms of the absolute inlet velocity C_1, remembering that $C_1^2 = W_1^2 - U_1^2$:

$$\text{NPSHR} = \frac{1}{\lambda^2}\frac{C_1^2}{2g} + \frac{U_1^2}{2g}\left\{\frac{1}{\lambda^2} + 1 - 2\sin\beta_1 \frac{1}{\lambda}\left[\left(\frac{C_1}{U_1}\right)^2 + 1\right]^{1/2}\right\}$$

Similar and much more detailed calculations of the separated flow region on the blade leading edges have been done, especially in conjunction with the development of inducers (Brennen 1994).

Example

Let us assume an inlet blade angle of $\beta_1 = 65°$ and a flow rate 40% higher than the design flow rate $Q/Q_0 = 1.4$. The tangent of the inlet flow angle is the ratio of the circumferential velocity U_1 and the meridional velocity C_1, and since the meridional velocity is proportional to the flow rate, it can be written

$$\tan \beta_{F1} = \frac{U_1}{C_1} = \left(\frac{Q_0}{Q}\right)\tan\beta_1$$

These values apply to the blade leading edge at the shroud where cavitation typically first appears, and assume a correct alignment of the flow velocity with the blade angle at design conditions. Similar calculations can be executed for other locations along the leading edge.

Therefore, the flow angle at the flow rate $Q = 1.4Q_0$ becomes $\beta_{F1} = 56.86°$. The quantities entering into the calculation of the velocity ratio $\lambda = W_1/W_s$ are $\cos \beta_{F1} = 0.54665$, $\cos(2\beta_{F1} - \beta_1) = 0.6080$, $\cos \beta_1 = 0.4226$.

Therefore, $\lambda = 0.67407$. Since $C_1/U_1 = 1/\tan \beta_{F1}$, the expression for the net positive suction head required becomes

$$\text{NPSHR} = 2.2 \frac{C_1^2}{2g} - 0.01144 \frac{U_1^2}{2g}$$

These results apply to the impeller inlet at the shroud and assume that the blade leading edge is radial, since the expression was derived from a two-dimensional model. In the case of a slanted leading edge, boundary layer development becomes more complex, and a leading edge vortex might appear. Also, some designers allow a slight positive incidence at design conditions.

An experimental correlation based directly on the inlet velocities has been published by Gongwer (1941). It is intended to predict the NPSHR value at correct incidence. Here the axial or meridional inlet velocity C_1 was equal to the average velocity calculated from the ratio of the flow rate and the inlet cross section. In reality, because of the curvature of the flow, the meridional velocity at the shroud, where cavitation first appears, is higher than the average value. There is also the issue of whether the blockage from blade thickness was taken into account. Consequently, this experimentally determined relationship may incorporate some effect of a velocity higher than the average meridional velocity that would typically have been encountered in the pumps from which the data were derived. Theoretically, at the design point the NPSHR would be expected to be equal to the velocity head of the local, meridional velocity

$$\text{NPSHR} = 1.485 \frac{C_1^2}{2g} + 0.085 \frac{U_1^2}{2g}$$

A similar expression has been published more recently (Turton 1994, p. 24) in the form

$$\text{NPSHR} = 1.8 \frac{C_1^2}{2g} + 0.23 \frac{U_1^2}{2g}$$

where C_1 and U_1 are in ft/sec. The Hydraulic Institute recently issued guidelines for rating centrifugal pumps while taking the available NPSHA into account (Hydraulic Institute 1998).

At a certain flow rate below the design flow rate, inlet recirculation may set in which completely alters the flow pattern along the leading edge. The two-dimensional leading-edge flow separation model loses its validity. Generally, less suction head is needed at reduced flow, and the danger of cavitation is reduced. However, a leading-edge vortex may appear with very low pressures at the centerline of the vortex, which may cause unexpected cavitation.

CAVITATION DAMAGE CORRELATION

Quantifying erosion damage estimates has recently received much attention (Hattori et al. 1998). A major research effort was initiated not too long ago by the Electric Power Research Institute to find quantitative measures of cavitation damage in centrifugal pumps (Guelich 1989 a,b). Detailed pump tests and a simultaneous visual observation of the actual cavitation bubbles led to the conclusion that cavitation damage could be correlated with the length of the cavitation bubble. The material removal rate E_R was found to be proportional to the following variables:

$$E_R = C_L \left(\frac{L_{cav}}{L_{cavR}}\right)^n (p_{oabs} - p_{vp})^3 R_m^{-2} \frac{a}{a_R} \left(\frac{\rho_{vp}}{\rho_{vpR}}\right)^{-0.44} \left(\frac{\alpha}{\alpha_R}\right)^{-0.36}$$

The coefficient of proportionality is $C_1 = 0.6 \times 10^{-6}$ in^3/lbf-sec (2.3×10^{-12} m^3/N·s) for the suction side and $C_1 = 0.3 \times 10^{-4}$ in^3/lbf-sec (1.1×10^{-10} m^3/N·s) for the pressure side, provided that the variables are normalized with the following reference values of the variables: cavity length $L_{cavR} = 0.4$ in. (1 cm), tensile strength of the blade material $R_m = 150,000$ psi (1000 N/mm^2), which was for ferritic steel in surface water, speed of sound in the vapor $a_R = 4888$ ft/sec (1490 m/s), density of the vapor $\rho_{vpR} = 0.00108$ lbm/ft^3 (0.0173 kg/m^3), gas content of water $\alpha_R = 24$ ppm, typical for surface water. The exponent n was 2.83 for the suction side and 2.6 for the pressure side.

In the study above the cavity length divided by the inlet diameter, L_{cav}/D_1, was found to generally decline from 0.2 at $\sigma = 0.2$ to 0.5, to 0.005 at $\sigma = 0.5$ to 2.0, depending on the flow rate as a fraction of the design flow rate. The blade material properties needed further corrective factors beyond the tensile strength of the reference ferritic material. Austenitic steels were found to be more resistant by a factor of 1.7. The weight loss varied from 0.00002 to 0.00004 lb/hr (0.0001 to 0.0002 N/h). Obviously, the data show large scatter, and the principal value of the correlation resides in the designer's ability to scale cavitation data derived from tests of existing pumps to new designs.

A correlation has been established (Sloteman 1995) between the difference of the cavitation number based on the available NPSHA value and the NPSH value at which a 3% head loss is observed, $\sigma_A - \sigma_{3\%}$, and the cavity length divided by the blade spacing at the inlet circumference, $L_{cav}/(\pi D_1/Z)$. Cavitation damage was also correlated with cavitation noise, which promises to be an important diagnostic tool (Guelich 1989 a,b; Guelich et al. 1993). However, this noise correlation does not enter into pump design.

INDUCERS

Cavitation on the impeller blades at the inlet can be avoided or mitigated if the pressure of the approaching fluid is raised by some means. In critical cases an

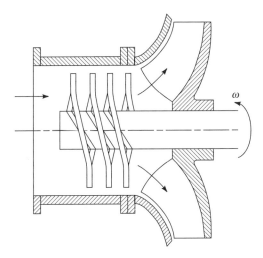

Figure 9.2 Inducer.

inducer ahead of the impeller inlet can achieve such a pressure rise (Fig. 9.2). Inducers make practical pump operation possible when the fluid actually boils or very nearly boils, as for example in cryogenic rocket pumps, aircraft fuel pumps (Rohatgi 1995), or when pumping some organic fluids which are near boiling in chemical processes. Government research on rocket technology and hydrofoils has imparted a great impetus to inducer research (Acosta 1973, Brennen 1994, Cooper 1967).

Inducers consist of axial flow rotors with spiral-shaped blades (Fig. 9.2). A design procedure can be found in a recent article by Rohatgi (1995). Designers use only three or four blades inclined 7 to 14° to the circumferential direction. The leading edge must be as sharp as possible. The axial or meridional velocity C_1 is kept low, and a slight positive incidence is preferred. Low axial velocities require large inlet cross sections, larger than the impeller inlet would normally have. Therefore, pumps using inducers and intended for very low suction heads must be custom designed. They need an exceptionally large inlet cross section for the inducer. The flow passage cross section decreases at the centrifugal impeller inlet to increase the meridional velocity and results in reasonable impeller inlet blade angles.

VIBRATION

Cavitation often appears together with noise and vibration (Brennen 1994). Two-phase flow of liquid and gas mixtures is well known for the potential of generating pressure fluctuations. However, since cavitation usually appears first at off-design conditions, in which flow separation and recirculation can exist even in the absence of cavitation, the causes and effects of vibration and cavitation remain difficult to separate. Flow separation, vortices, and mixing of the fluid in the impeller are intimately connected with the cavitation process, but also exist in the absence of cavitation.

Pressure fluctuations and vibration have an important effect on rotor dynamics, on the sizing of the pump bearings, and on the corresponding bearing life. The magnitude and duration of bearing loads resulting from vibrations cannot be estimated accurately. An additional safety margin must be added to the anticipated bearing load when frequent operation at off-design conditions are anticipated or under cavitation. Violent vibrations might even bring about a catastrophic bearing overload. At present in design engineering calculations, the only practical approach to account for such upset conditions is to add a sufficient safety margin. Standards of the Hydraulic Institute, ISO, ASME, and API prescribe permissible vibration levels. Rotating equipment and bearing manufacturers often offer guidelines to estimate the service factors appropriate for certain applications. Vibrations can be measured during pump tests, and their anticipated long-term effect on bearing life can be estimated.

REFERENCES

Acosta, A. J. (1973): Hydrofoils and Hydrofoil Craft, *Annual Review of Fluid Mechanics*, Vol. 5, pp. 161–184.

Arndt, R. E. A. (1981): Cavitation in Fluid Machinery and Hydraulic Structures, *Annual Review of Fluid Mechanics*, Vol. 13, pp. 273–328.

Avellan, F., et al. (1995): *La cavitation: mécanismes physiques et aspects industrielles*, Presses Universitaires de Grenoble, Grenobles, France.

Brennen, C. E. (1994): *Hydrodynamics of Pumps*, Oxford University Press, New York, 1994.

Cooper, P. (1967): Analysis of Single and Two-Phase Flow in Turbo-Pump Inducers, *ASME Journal of Engineering for Power*, October, pp. 577–588.

Cornell, W. G. (1962): The Stall Performance of Cascades, *Proceedings of the 4th U.S. National Congress on Applied Mechanics*, ASME, New York, pp. 1291–1299.

Gongwer, C. A. (1941): A Theory of Cavitation Flow in Centrifugal-Pump Impellers, *ASME Transactions*, January, Vol. 63, pp. 29–40.

Guelich, J. F. (1989): *Guidelines for Prevention of Cavitation in Centrifugal Feedpumps*, EPRI Paper GS-6398, November.

Guelich, J. F. (1989): Cavitation Erosion in Centrifugal Pumps, *Chemical Engineering Progress*, November, pp. 68–73.

Guelich, J. F., Cother, A., Martens, H. J. (1993): Cavitation Noise in Jet Cavitation Tests and Pumps, in *Pumping Machinery*, ASME FED Vol. 154, pp. 1–6.

Hattori, S., Mori, H., Okada, T. (1998): Quantitative Evaluation of Cavitation Erosion, *ASME Journal of Fluids Engineering*, March, pp. 179–186.

Hydraulic Institute (1998): *Centrifugal and Vertical Pumps for NPSH Margin*, ANSI/HI 9.6.1, HI, Parsippany, N. J.

La Houille Blanche (1988): No. 7/8.

La Houille Blanche (1992): No. 7/8.

Rohatgi, U. S. (1995): Sizing of Aircraft Fuel Pump, *ASME Journal of Fluids Engineering*, June, pp. 298–302.

Salemann, V. (1959): Cavitation and NPSH Requirements of Various Liquids, *ASME Journal of Basic Engineering*, June, pp. 167–180.

Sloteman, D. P. (1995): Avoiding Cavitation in the Suction Stage of High Energy Pumps, *World Pumps*, September, pp. 40–48.

Stepanoff, A. J. (1957): *Centrifugal and Axial Flow Pumps*, Wiley, New York. Reprinted by Krieger Publishing, Malabar, FLa., 1993, pp. 250–253.

Sulzer Brothers Ltd. (1989): *Centrifugal Pump Handbook*, Elsevier Applied Science, New York, p. 28.

Turton, R. K. (1994): *Rotodynamic Pump Design*, Cambridge University Press, New York, 1994.

Vlaming, D. J. (1989): Optimum Impeller Inlet Geometry for Minimum NPSH Requirement for Centrifugal Pumps, in *Pumping Machinery*, ASME FED Vol. 81, pp. 25–28.

10

PERFORMANCE CALCULATION

CALCULATION PROCEDURE

A calculation procedure to estimate the theoretical performance of a pump is an indispensable tool in pump design. Performance needs to be known, not only at the rated, best efficiency point, but also off design. Pump specifications often impose special requirements, such as head at shutoff, maximum power demand, rate of head rise to assure stability, and so on. A good pump design process requires trial-and-error iteration, a check on anticipated performance with a trial geometry, and progressive approximation to the optimal design configuration. The best hydraulic design does not necessarily correspond to the best commercial pump product. Compromises are unavoidable.

Excellent pump performance calculation procedures have been developed and published (Balje 1981, Church 1944, Guelich 1999, Hamkins 1984, Ida 1979, Jansen and Sunderland 1990, Peng and Jenkins 1983, Pfleiderer 1961, Stepanoff 1965, Traupel 1988, Turton 1994), and several computer programs are commercially available. A computer program developed by NASA and available from COSMIC (Veres 1994) relies heavily on empirical correlations compiled from rocket pump data and does not identify the various loss mechanisms. Generally, the more accurate and detailed these calculations are, the greater the number of input variables needed: not only the desired head, flow rate, and rotational speed, but also the details of the geometrical description of the impeller and housing. The labor and cost required to prepare and enter the input variables limits the practical use of computer programs for simple, inexpensive pumps.

A simple and fast calculation procedure with minimal input requirements, is presented here (Tuzson and Iseppon 1997). The mathematical expressions,

122 PERFORMANCE CALCULATION

programmed in the computer code LOSS3 and shown in the listing in the Appendix, are presented on the following pages. Flow conditions are calculated on an average streamline through the pump. It should be particularly helpful for modifying or improving existing designs, provided that it is first adjusted to match the performance of a similar, existing pump. In the process of matching the performance of an existing pump, the various loss coefficients are determined. Many of these loss coefficients depend on the existing manufacturing practices of the pump manufacturer and would be expected to remain the same for new designs. Such factors might, for example, be the machining and casting surface finishes in use.

Since FORTRAN listings do not permit variables with subscripts, and to minimize the need for two sets of symbols for the same variables, the equations in this chapter are often presented in terms of the symbols used in the computer program. In some instances the use of two symbols for the same variable cannot be avoided. However, symbols are clearly defined, and a full nomenclature is given with the computer program listing in the Appendix. The expressions in the program listing include conversion factors since the code was intended for numerical calculations in the English system. Use with the SI system would require a review and modification of these conversion factors. As an alternative, a conversion from SI to English, and vice versa, can be added at the beginning and end of the program.

Designers can easily implement and use this calculation procedure on simple personal computers. The Appendix contains a three-page FORTRAN listing, LOSS3. The mathematical expressions of the listing can also be incorporated into larger, proprietary computer programs. Since the procedure and program listing are simple, users can and should understand the physical phenomena behind the calculations. The calculation can be executed by hand, but the computer can do it faster. Every calculation step can be checked by hand. If the results do not make sense, reasonable adjustments should be made in the input. The designer and user should use practical judgment to override, if necessary, the course of action suggested by the program. It should be kept in mind that pump efficiency depends not only on the hydraulic design but also on the mechanical inefficiency, for example.

The effect of individual variables, such as number of impeller blades, inlet diameter, and impeller exit width (Fig. 10.1) on pump performance is complex and often contradictory, depending on the value of other variables. Their effect on performance is difficult to predict from memory or from empirical rules, because of the complexity of their interdependence. A simple computer program can quickly show, if not the exact numerical magnitude, at least the trend of their effect. This calculation procedure is intended for radial or mixed flow pumps with return passages or volutes, in a range of specific speeds from 500 to 5000 calculated in English units, 0.2 to 2.0 in dimensionless units, and flow rates from 100 to 5000 gpm, equivalent to 0.006 to 0.3 m^3/s.

Depending on the particular case, the calculated head and efficiency might approach the measured value to within 2%, which is usually adequate since pump test data are rarely more accurate or reproducible than 1%. The greatest accuracy

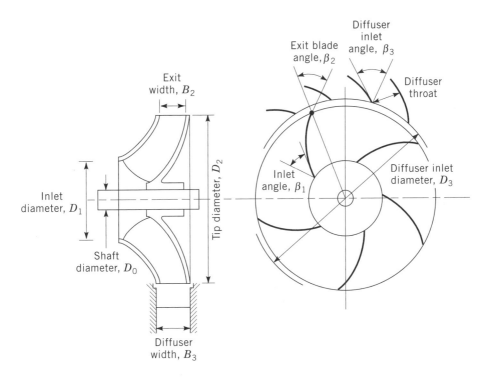

Figure 10.1 Nomenclature of impeller and diffuser geometry.

can be expected near best efficiency, where the losses are the smallest. The basic concept of the calculation consists of estimating the energy imparted by the pump impeller to the fluid, the theoretical head, and of subtracting all head losses to arrive at the net output head for a given flow rate. To calculate efficiency, the power absorbed by disk friction, the bearings, and inlet recirculation are added to the hydraulic power demand of the impeller. The flow rate is adjusted for leakage.

The various flow phenomena corresponding to the losses and velocities calculated in the computer program are discussed in Chapter 7. The mathematical expressions used in the program are presented in the present chapter. The computer program requires input of the design flow rate, rotational speed, geometrical variables of the pump, and several loss coefficients, as shown on the printout of the design example. The head, efficiency, losses, leakage, and several velocities are printed out for six selected flow rates across the operating range. The Appendix lists the nomenclature and specific instructions for executing the computer program.

IMPELLER WORK INPUT

The energy imparted by the impeller to the fluid, the theoretical hydraulic head H_{th}, designated DHTH in the computer program, is calculated from Euler's

124 PERFORMANCE CALCULATION

equation, which corresponds to the conservation of angular momentum. No prerotation is assumed here. When prerotation is present, the flow approaching the impeller rotates in the direction of, or opposite to, the sense of rotation of the impeller. The angular momentum of the fluid entering the impeller would then have to be subtracted from, or added to, the angular momentum of the fluid leaving the impeller, in order to calculate the theoretical hydraulic head. Without prerotation the theoretical hydraulic head is given by the expression

$$H_{th} = \frac{U_2 C_{t2}}{g} = \frac{U_2^2 \sigma - U_2 W_{m2} \tan \beta_2}{g}$$

where

$U_2 = (2\pi/60)N = $ tip speed
$N = $ rpm $= $ rotational speed
$W_{m2} = Q/2\pi R_2 B_2 = $ radial or meridional velocity
$R_2 = $ impeller exit radius
$B_2 = $ effective impeller exit width
$\sigma = $ slip coefficient according to Wiesner
$C_{t2} = $ absolute tangential velocity at impeller exit
$\beta_2 = $ exit blade angle measured from radial or meridional, in direction opposite to the sense of rotation

If significant blockage, due to blades, boundary layers, or separation is anticipated at the impeller exit, the effective impeller width B_2 can be made somewhat smaller (but never larger) than the geometrical width. In mixed flow impellers the meridional velocity W_{m2} is inclined toward the rotational axis, and the impeller width B_2 is to be taken perpendicular to the direction of the velocity even if the exit leading edge is slanted. The exit angle β_2 should be measured in the plane of the exit velocity vector W_2 and its meridional projection W_{m2}. If the housing restricts the impeller exit, as in some vertical, multistage pumps, the slip coefficient can become smaller than the value given by Wiesner's correlation, and the head will be reduced.

FLOW LOSSES

Incidence losses, DQIN12, at the pump inlet are calculated by assuming a leading-edge separation, and a sudden expansion loss, when the separated flow mixes. These flow patterns at incidence, illustrated in Fig. 10.2, have been described in previous chapters. Calculation of the extent of the separated region and the corresponding velocity increase follows a potential flow model, solved by conformal transformation (Cornell 1962). The inlet flow angle β_{F1}, BEI in the listing, is calculated from $\tan \beta_{F1} = U_1/C_1$, and the blade angle is β_1, BEB1 in

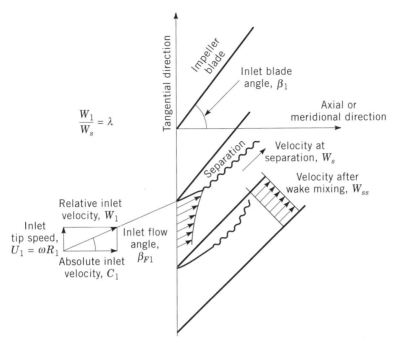

Figure 10.2 Impeller inlet incidence and separation.

the listing, both measured from the meridional direction. The following equation of the potential flow calculation determines the ratio λ of inlet velocity W_1 to the velocity at separation W_s:

$$\lambda = \frac{W_1}{W_s} = \frac{\cos \beta_{F1}}{\cos(2\beta_{f1} - \beta_1)} - \left\{ \frac{(\cos \beta_{F1})^2 - \cos \beta_1 \cos(2\beta_{F1} - \beta_1)}{[\cos(2\beta_{F1} - \beta_1)]^2} \right\}^{1/2}$$

Assuming a sudden expansion loss beyond the separated region, and substituting $W_{ss} = W_1 \cos \beta_{F1} / \cos \beta_1$, the expression for the head loss becomes

$$\text{DQIN12} = \frac{W_s^2}{2g} \left(1 - \frac{W_{ss}}{W_s} \right)^2 = \frac{W_1^2}{2g} \left(\frac{1}{\lambda} \right)^2 \left(1 - \frac{\lambda \cos \beta_{F1}}{\cos \beta_1} \right)^2$$

Here W_{ss} stands for the velocity far downstream from the separated region, where the separated flow has been mixed and the velocity becomes uniform.

The expression above, which is programmed into the performance calculation computer program, was derived for a broad range of blade cascade configurations. A simpler form will give essentially identical results. It assumes that the loss is proportional to the square of the difference between the tangential component of the inlet velocity C_{t1} and the circumferential velocity U_1:

$$\text{DQIN12} = \frac{(U_1 - C_{t1})^2}{2g} = \frac{(U_1 - C_1 \tan \beta_1)^2}{2g}$$

Therefore, the inlet incidence loss calculation does not include any empirical loss coefficient. The calculation assumes a radial leading edge and an axial inlet velocity. In the case of a slanted leading edge, the inlet blade angle is to be adjusted artificially to achieve minimum incidence loss at the design flow rate. For calculating accurate inlet flow conditions, a separate, preferably three-dimensional calculation procedure should be used. The computer program INLET in the Appendix offers a simple approximation.

The wall friction or skin friction losses (DQSF12 and DQSF23) in the impeller and diffuser or volute follow the standard pipe friction model (Hydraulic Institute 1998). Since the flow passage cross sections are irregular, a hydraulic radius and average flow velocities are used. The friction coefficient can be adjusted but has a default value of 0.005. The hydraulic radius of the impeller blade passages is obtained by dividing the passage cross sectional area by half of the circumference. The geometrical measurements are taken at the impeller exit: B, the blade height, becomes the passage width, and the circumference, πD2, divided by the number of blades, XNB, and multiplied by the cosine of the blade angle at the exit, BEB2, becomes the distance perpendicular to the blade surface:

$$\text{DHYD12} = \text{B}(\pi\text{D2}/\text{XNB})\cos\text{BEB2}/(\text{B} + (\pi\text{D2}/\text{XNB})\cos\text{BEB2})$$

The blade passage length is taken to be $(\text{D2} - \text{D1})/(2\cos\text{BEB2})$. Finally, assuming an average relative flow velocity $(W2 + W1)/2$, the friction head loss becomes

$$\text{DQSF12} = \text{CFS}[(\text{D2} - \text{D1})/(2\cos\text{BEB2})](1/\text{DHYD12})(W2 + W1)^2/4g$$

A diffusion loss (DQDIF) needs to be taken into account, since separation invariably appears in the impeller at some point. The program assumes that when the ratio of the relative velocity at the inlet W_1 and the outlet W_2 exceeds a value of 1.4, a portion of the velocity head difference is lost:

$$\text{DQDIF} = \frac{0.25 W_1^2}{2g}$$

if $W_1/W_2 > 1.4$.

Volute head loss (DQIN23) results from a mismatch of the velocity leaving the impeller and the velocity in the volute throat. If the velocity approaching the volute throat is larger than the velocity at the throat, the velocity head difference is lost (Gopalakrishnan and Lorett 1986, Worsted 1963). The program calculates the velocity approaching the volute throat by assuming that the velocity leaving the impeller C_{t2} decreases in proportion to the radius because of the conservation of angular momentum: $C_3 = C_{t2}(D_2/D_3)$. The volute throat velocity C_{Q3} is calculated from the flow rate and the volute throat cross-sectional area:

$$\text{DQIN23} = 0.8 \frac{(C_3^2 - C_{Q3}^2)}{2g}$$

To calculate the diffuser loss (DQVD) the program estimates the pressure recovery coefficient based on the area ratio and applies an adjustable loss coefficient CVD. In pumps the diffuser accounts for the greatest head loss, much influenced by the detailed design of the diffuser. Therefore, estimating diffuser losses is particularly difficult (Reneau et al. 1967).

$$\text{DQVD} = \text{CVD} \left(\frac{C_{Q3}^2}{2g} \right)$$

Some designers assume, as a rule of thumb, that half of the diffuser inlet velocity head is lost.

The disk friction loss (DFH) calculation follows the well-known expression, the default value of the loss coefficient CDF being 0.005 (Nece and Daily 1960). The disk friction power is divided by the flow rate and appears as a parasitic head, to be added to the theoretical head when the shaft power demand is calculated.

$$\text{DFH} = \frac{\text{CDF} \cdot \rho \omega^3 (D/2)^5}{Q}$$

A parasitic power demand also results from the appearance of inlet recirculation (DRECH) (Tuzson 1983). The program uses an expression of the type

$$\text{power of recirculation} = \text{CREC} \cdot \omega^3 D_1^2 \left(1 - \frac{Q}{Q_0}\right)^{2.5}$$

where D_1 is the inlet diameter, Q_0 is the design flow rate.

Unfortunately, the recirculation loss coefficient depends on the piping configuration upstream of the pump in addition to the geometrical details of the inlet, and therefore is difficult to estimate. Impellers with relatively large inlet diameters (usually encountered in high-specific-speed pumps) are the most likely to recirculate. The program contains a default value of 0.005 for the loss coefficient, but verification of such a value on similar pumps is advisable. The parasitic power of recirculation is also divided by the volume flow rate, like the disk friction power, in order to be converted into a parasitic head.

The leakage flow rate (QL) across the front wear ring clearance is calculated from a simple orifice equation (Bilgen et al. 1973), the cross-sectional area of the flow being the clearance width—here taken to be a fixed at 5/1000 in. (0.125 mm)—multiplied by the inlet circumference. The head difference across the clearance is assumed to be the impeller exit total or theoretical head DHTH, from which one quarter of the velocity head of the tip speed ($\frac{1}{4}U_2^2/2g$) is deducted. This approach implies the assumption that the fluid between the impeller shroud and the stationary housing rotates with half the impeller speed. A contraction coefficient of 0.8 is used.

$$QL = 0.8(0.005\pi D1)\left(DHTH - \frac{U_2^2}{8g}\right)$$

The leakage flow rate is deducted from the impeller flow rate when the efficiency is calculated. Usually, it only amounts to 1 or 2% of the total flow rate. The calculated value is very approximate and serves to give some idea as to whether leakage losses are significant. The wear ring clearance width depends on the manufacturing tolerances and becomes relatively small in large pumps. Shaft deflection must also be taken into account when the least possible clearance width is estimated.

Finally, the pump head is calculated by subtracting from the theoretical head all the flow losses:

$$DH = DTH - (DQIN12 + DQSF12 + DQSF23 + DQIN23 + DQDIF + DQSF34 + DQVD)$$

The efficiency is calculated by dividing the pump head by the theoretical head, to which the disk friction head and the recirculation head have been added, and multiplying this ratio by the ratio of the output flow rate and the sum of the output flow and the leakage flow rate:

$$EFF = 100(DH/(DHTH + DFH + DRECH))(Q/(Q + QL))$$

CALIBRATION ON TEST DATA

The performance calculation computer program will not give satisfactory results unless it has been calibrated on a similar pump. The loss coefficients depend on the design and manufacturing practices of the pump manufacturer. For example, the skin friction coefficient will depend on the machining and casting surface finishes. Detailed information on typical surface finishes can be found in an article by Osterwalder and Hippe (1982). However, the most practical method to establish the specific values of these loss coefficients is to compare and adjust the calculation results to the tested performance of a similar, existing pump. Since the design and manufacturing procedures of the new pump are likely to be the same as those of the old pump, the loss coefficients are also likely to be the same.

The calibration process, the adjustment of the calculation results by changing the loss coefficients, also teaches the user the effect of each loss mechanism on the overall performance of the pump. An intuitive feel develops for recognizing specific features of the pump performance just by looking at the head–flow and efficiency curves. Low efficiency at certain flow rates, or features of the head–flow curve, can be connected with specific loss mechanisms. The difference between the performance details of low- and high-specific-speed pumps becomes apparent.

It is indispensable to first estimate correctly the energy input by the impeller. Subsequent losses can only dissipate, can never add, energy. The first step in matching calculation to tested performance consists of verifying the work input, the head H divided by the efficiency η of the pump (Fig. 10.3). The work input head multiplied by the flow rate corresponds to the power demand of the pump. As shown on the head versus flow rate plot of Fig. 10.3, the curve of the work input becomes parallel and almost identical to the theoretical head line at large flow rates, beyond the design flow rate. The work input head consists of the sum of the theoretical head (DHTH), the recirculation head (DRECH), and the disk friction head (DFH), which also includes mechanical losses, such as bearing losses. The head corresponding to these various losses decreases rapidly with increasing flow rate. Therefore, at high flow rates the work input becomes practically identical with the theoretical head. The theoretical head is sometimes called the *ideal hydraulic head*, which is the energy per unit volume that the impeller imparts to the fluid.

Figure 10.3 also shows that the theoretical head follows a straight line, which intersects the vertical axis at shut off, at a value equal to $U_2^2 \sigma/g$ which depends only on the tip speed of the impeller. The slip remains the same over the entire flow range for a given blade angle and number of blades and is relatively insensitive to small changes. The rotational speed and impeller diameter, which determine the tip speed, are generally known and are often given in pump catalogs.

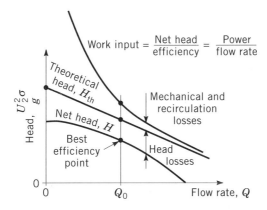

Figure 10.3 Calculated head–flow curve of a centrifugal pump.

130 PERFORMANCE CALCULATION

The impeller exit blade angle β_2 and the exit width B_2 determine the slope of the line corresponding to the theoretical head, as can be seen from its mathematical expression. Pivoting around the intersection point at shutoff, and tangent to the work input at high flow rates, the slope of the line is easily found and adjusted. There is some justification for deviating from the exact, geometrical measurements of the variables. The average exit blade angle is difficult to measure and does not always correspond to the value shown on drawings. The effective impeller exit width can be made smaller (but never larger) if there is reason to suspect some blockage, due to blockage of blade thickness, boundary layers, or flow separation in the impeller.

As to the effect of losses (Fig. 10.4), skin friction and diffuser losses increase with the square of the flow rate and predominate at high flow rates. Therefore, these loss coefficients should be adjusted when the calculated head deviates from the experimental head at high flow rates. The volute loss depends on the impeller exit velocity, which increases with decreasing flow rate. Therefore, the volute loss coefficient should be adjusted when the calculated head deviates from the experimental head at low flow rates. Incidence losses increase on either side of the design flow rate but have no adjustable loss coefficient. The computer program assumes a radial leading edge. The incidence losses are often fictitious, and the inlet blade angle should be adjusted to achieve minimal losses at the design flow rate—unless a bad incidence is suspected at the design flow rate—to make the calculation agree with the test data. The disk friction and recirculation losses, which also include the mechanical losses of the bearings, affect only the efficiency, not the head.

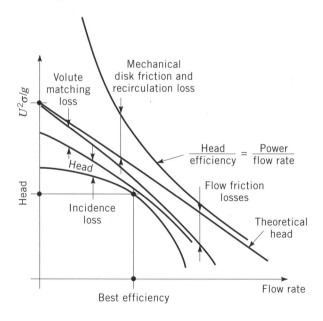

Figure 10.4 Pump head losses.

RESULTS

The printout of the computer program shows the flow rate, head, efficiency, individual losses, and velocities at various locations in the pump. Unfortunately, testing only informs about the overall performance of the pump. Measurement of the flow pattern and velocities inside the pump, especially in the impeller, presents great difficulties and would be prohibitively expensive in the case of most pumps. If a satisfactory match is achieved between tested and calculated performance, there is reason to believe not only that the losses are being assessed correctly but that the calculated average velocities also approximate well the real flow velocities. Since high velocities usually signal high losses, the location of potential improvements can be identified. Knowing the local velocities provides additional information toward an understanding of the behavior of the flow inside the pump.

The design and development of pumps with exceptional performance requires alternating steps of experiment and analysis. An execution of the performance calculation during or immediately following an experiment can reveal the details of shortcomings and suggest opportunities for further improvement in the next iteration cycle.

REFERENCES

Balje, O. E. (1981): *Turbomachines*, Wiley, New York.

Bilgen, E., Boulos, R., Akgungor, A. C. (1973): Leakage and Frictional Characteristics of Turbulent Helical Flow in Fine Clearance, *ASME Journal of Fluids Engineering*, December, pp. 493–497.

Church, G. (1944): *Centrifugal Pumps and Blowers*, John Wiley, New York, 1944.

Cornell, W. G. (1962): The Stall Performance of Cascades, *Proceedings of the 4th U.S. National Congress on Applied Mechanics*, ASME, New York, pp. 1291–1299.

Gopalakrishnan, S., Lorett, J. A. (1986): Interaction Between Impeller and Volute of Pumps at Off-Design Conditions, *ASME Journal of Fluids Engineering*, March, pp. 12–18.

Guelich, J. F. (1999): *Kreiselpumpen*, Springer-Verlag, New York.

Hamkins, C. (1984): *Correlation of One-Dimensional Centrifugal Pump Performance Analysis Method*, ASME Paper 84-WAFM-10.

Hydraulic Institute (1998): *Pipe Loss Standards*, HI, Parsippary, N.J.

Ida, T. (1979): Analysis of Scale Effects on Centrifugal Pumps, *Science Reports of the Research Institute for Engineering*, Kanagawa University, Kanagawa, Japan, August.

Jansen, W., Sunderland, P. B. (1990): Off-Design Performance Prediction of Centrifugal Pumps, in *Fluid Machinery Components*, ASME FED Vol. 101, pp. 1–9.

Nece, R. E., Daily, J. W. (1960): Roughness Effects on Frictional Resistance of Enclosed Rotating Disks, *ASME Journal of Basic Engineering*, September, pp. 553–562.

Osterwalder, J., Hippe, L. (1982): Studies on Efficiency Scaling Process of Series Pumps, *IAHR Journal of Hydraulic Research*, Vol. 20, No. 2, pp. 175–199.

Peng, W. W., Jenkins, P. E. (1983): Hydraulic Analysis on Component Losses of Centrifugal Pumps, in *Performance Characteristics of Hydraulic Turbines and Pumps*, ASME FED Vol. 6, pp. 121–125.

Pfleiderer, C. (1961): *Die Kreiselpumpen*, Springer-Verlag, Berlin, 1961.

Reneau, L. R., Johnston, J. P., Kline, S. J. (1967): Performance and Design of Straight, Two-Dimensional Diffusers, *ASME Journal of Basic Engineering*, March, p. 141.

Stepanoff, A. J. (1965): *Centrifugal and Axial Flow Pumps*, Wiley, New York. Reprinted by Krieger Publishing, Malabar, Fl., 1993.

Traupel, W. (1988): *Thermische Turbomaschinen*, Springer-Verlag, Berlin, 1988.

Turton, R. K. (1994): *Rotodynamic Pump Design*, Cambridge University Press, New York, 1994.

Tuzson, J. (1983): Inlet Recirculation in Centrifugal Pumps, in *Performance Characteristics of Hydraulic Turbines and Pumps*, ASME FED Vol. 6, pp. 195–200.

Tuzson, J. and Iseppon, A. (1997): Centrifugal Pump Design Teaching Tool, *ASME Pumping Machinery Symposium*, FEDSM97-3370.

Veres, J. P. (1994): *Centrifugal and Axial Pump Design and Off-Design Performance Prediction*, NASA Technical Memorandum 106745, COSMIC Program LEW-16173.

Wiesner, F. J. (1967): A Review of Slip Factors for Centrifugal Impellers, *ASME Journal of Fluids Engineering*, October, pp. 558–572.

Worsted, R. C. (1963): The Flow in Volutes, *Proceedings of the Institution of Mechanical Engineers*, Vol. 177, No. 31, pp. 843–875.

11

PUMP DESIGN PROCEDURE

PUBLISHED DESIGN PROCEDURES

Several pump design procedures are described in the literature, including methods covered in the books of Balje, Church, Guelich, Pfleiderer, Stepanoff, Traupel, and Turton cited in Chapter 10. Others have been published in engineering journals (e.g., Peck 1968, Salisbury 1983, Spring 1984) and in government reports (e.g., Oak Ridge 1972). These methods are mentioned here for comparison with the procedure offered in this book. These design procedures concern themselves primarily with design point operation. Without a calculation to predict performance across the operating range, the loop of design iteration cannot be closed. Development would then proceed immediately to expensive fabrication and testing at considerable risk. The design procedure advocated here relies on the performance calculation computer program to verify that design specifications have been reached.

CHOICE OF IMPELLER DIAMETER, EXIT WIDTH, AND EXIT BLADE ANGLE

The design procedure presented here will lead to a first, trial pump configuration. It will choose the values of the input variables needed by the performance calculation computer program. Execution of the performance calculation will show whether the desired characteristics and efficiency have been attained. Modifications can then be tested to optimize the design. After a satisfactory overall design has been achieved, detailed design of the components—the impeller blading, volute configuration, or return passage vanes—can proceed.

In most cases the rotational speed N (rpm), design flow rate Q_0, and head H_0 will be prescribed. Electric motor drive prescribes standard motor speeds. Often, the maximum size of the impeller diameter D_2 is limited. Since the head can hardly exceed U_2^2/g, a fast check can be made to see if a reasonable impeller size can achieve the desired head or if several stages will be required [$g = 32.2$ ft/sec^2(9.8 m/s^2)]. The tip speed U_2 is calculated from

$$U_2 = \left(\frac{N 2\pi}{60}\right)\frac{D_2}{2}$$

Note that in English units the diameter is usually in inches and the tip speed is desired in ft/s, which introduces a conversion factor of 12.

To eliminate the effect of the losses for the time being, a target efficiency η will be assumed. The pump specifications often prescribe such an efficiency target. The prescribed efficiency can be checked against the efficiency derived from the plot in Fig. 11.1, which shows typical, industrywide efficiencies as a function of flow rate and specific speed. The Hydraulic Institute (1994) also offers an efficiency prediction method. Since the design head, flow rate, and speed are given, the specific speed can be calculated. Desired efficiencies higher than those given by the plot would not be easy to achieve, and would require an extensive combined analytical and experimental development process.

By dividing the given design head H_0 by the efficiency η, the theoretical head H_{th} can be calculated, which does not depend on the losses, and which determines how much energy the impeller has to impart to the fluid. To facilitate the choice of subsequent variables, the dimensionless plot of Fig. 11.2 is introduced, which shows the theoretical head, represented by the head coefficient ψ, against the flow rate, represented by the flow coefficient ϕ. These coefficients are related to the head rise across the pump H, flow rate Q, efficiency η, tip speed U_2, diameter D_2, and exit width B_2 of the pump by the expressions

$$\psi = \frac{gH}{\eta U_2^2}$$

$$\phi = \frac{Q}{\pi D_2 B_2 U_2}$$

Care must be taken when comparing these head and flow coefficients with those appearing in other publications, because some may be defined differently. Sometimes head coefficients do not take the efficiency into account and are based only on the net head rise of the pump. Sometimes, the head is normalized with $U^2/2g$ instead of U^2/g as is done above.

The definitions used here are chosen because they connect the theoretical head or energy input of the pump directly to the absolute tangential velocity at the

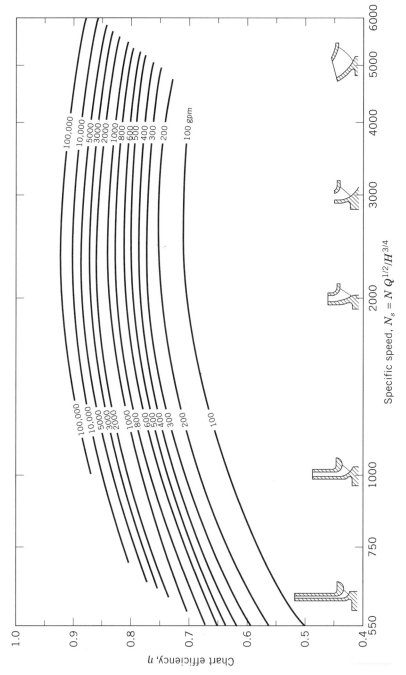

Figure 11.1 Chart efficiency of pumps versus specific speed. (From Oak Ridge 1972.)

136 PUMP DESIGN PROCEDURE

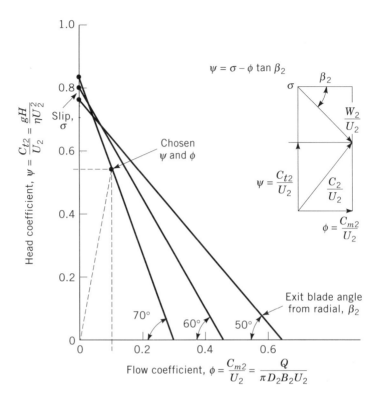

Figure 11.2 Head and flow coefficient diagram. The slip coefficient shown is for six blades.

impeller exit and the flow rate with the radial velocity at the impeller exit. The dimensionless head coefficient corresponds to the absolute tangential velocity at the impeller exit C_{t2} divided by the tip speed U_2. The dimensionless flow coefficient corresponds to the radial velocity at the impeller exit C_{r2} divided by the tip speed U_2:

$$\text{head coefficient} = \psi = \frac{C_{t2}}{U_2}$$

$$\text{flow coefficient} = \phi = \frac{C_{r2}}{U_2}$$

The plot shown in Fig. 11.2 actually reproduces the velocity diagram at the impeller exit in a dimensionless form. The tangential direction is along the vertical axis, and the meridional direction along the horizontal. As the flow rate changes, the operating point of the pump moves along the straight line of constant angle β_2:

$$\psi = \sigma - \phi \tan \beta_2$$

CHOICE OF IMPELLER DIAMETER, EXIT WIDTH, AND EXIT BLADE ANGLE

The velocities, C_{t2} proportional to the theoretical head, and C_{m2} proportional to the flow rate, fully determine the pump performance.

Several empirical correlations have been recommended in the literature between the head coefficient and the specific speed. According to these the head coefficient would generally decline, approximately linearly, from about $\psi = 0.75$ at $N_s = 1000$ to $\psi = 0.45$ at $N_s = 4000$ (Rey et al. 1982). Evidently, the head coefficient cannot be larger than the slip coefficient.

The expression of the theoretical head above shows immediately whether the desired head can be achieved with the tip speed and the head coefficient chosen. The exact impeller diameter D_2 can then be chosen to give the necessary tip speed U_2:

$$\text{theoretical head} = H_{th} = \frac{H_0}{\eta} = \frac{U_2 C_{t2}}{g} = \frac{\psi U_2^2}{g}$$

It should be mentioned that some pump designers assume a prerotation in the flow approaching the impeller (Spring 1983). Then the inlet velocity would have a tangential component C_{t1} and the angular momentum would not be zero at the inlet. The corresponding head $C_{t1} U_1 / g$ would then have to be deducted from the theoretical head. No such prerotation will be assumed here.

The expression of the flow rate determines the effective impeller exit width B_2. The geometrical impeller width can be made a few percent larger to account for blockage from the blades, from boundary layers, or from separation:

$$\text{flow rate} = Q = (\pi D_2 B_2) C_{r2} = (\pi D_2 B_2) U_2 \phi$$

$$B_2 = \frac{Q}{\pi D_2 U_2 \phi}$$

Consistent units must be used:

$$Q(\text{in}^3/\text{sec}) = Q(\text{gpm})(231 \text{ in}^3/\text{gall})/(60 \text{ sec}/\text{min})$$

Typical values for the head coefficient are $\psi = 0.4$ to 0.7, and for the flow coefficient are $\phi = 0.05$ to 0.2. For high-specific-speed pumps, lower head coefficients and higher flow coefficients are appropriate.

The choice of the head and flow coefficients from the plot of Fig. 11.2 also determines the impeller exit blade angle β_2, which in a typical pump is 68° from radial or meridional and 22° from tangential. It corresponds to the angle of the inclined, straight line, which represents the operating line of the pump. The line intersects the vertical axis at a value that is numerically equal to the slip coefficient. If the number of impeller blades is assumed—for example, six—a slip coefficient can be calculated from Wiesner's correlation, and the intersection of the operating line and the axis on the plot of Fig. 11.2 can be checked or modified. The line in the plot reproduces the expression for calculating the absolute tangential velocity at the impeller exit in a dimensionless form:

$$C_{t2} = U_2 \sigma - C_{m2} \tan \beta_2$$

$$\frac{C_{t2}}{U_2} = \psi = \sigma - \phi \tan \beta_2$$

The number of impeller blades can be selected on a trial basis, usually six, give or take one. The effect of more or fewer blades can be explored easily with the performance calculation computer program. More blades guide the flow better, increase the slip coefficient, and therefore increase the head somewhat. On the other hand, friction losses are increased, and the blockage produced by the blade thickness at the inlet could become excessive.

IMPELLER INLET DIAMETER

The value of the impeller inlet diameter D_1 is usually selected to minimize the inlet relative velocity. Low velocities favor low losses (Vlaming 1989). Low inlet relative velocities minimize the diffusion, the ratio of inlet to exit velocity W_1/W_2, in the impeller, which may lead to flow separation if excessive.

Theoretically, an optimal inlet diameter exists, as a trade-off between two opposing trends. The vectorial sum of the circumferential velocity U_1 and the axial velocity at design conditions C_1 adds up to the relative inlet velocity W_1. (It is assumed here that no prerotation exists.)

$$W_1^2 = U_1^2 + C_1^2$$

$$U_1 = \frac{\omega D_1}{2}$$

$$C_1 = \frac{Q}{\pi D_1^2/4}$$

Assuming a radial leading edge and an axial inlet velocity without swirl, the circumferential velocity U_1 at the shroud increases with the inlet diameter, while the axial inlet velocity C_1 decreases with the square of the inlet diameter for a given flow rate. The inlet velocity C_1 is given by the flow rate divided by the inlet cross-sectional area. An allowance has to be made for blockage from the blade thicknesses at the inlet, which is usually more critical than at the impeller exit. These opposing trends balance and make the inlet relative velocity a minimum, for a certain optimal inlet diameter

$$D_1 = \left(\frac{8\sqrt{2}}{\pi}\right)^{1/3} \left(\frac{Q}{\omega}\right)^{1/3} = 1.533 \left(\frac{Q}{\omega}\right)^{1/3}$$

radial

This optimal inlet diameter corresponds to an inlet blade angle β_1 of 54.73° from the axial direction, or 35.26° from the tangential direction, at the shroud. The rotational speed U_1 is then $(2)^{1/2} = 1.414$ times the inlet velocity C_1. The relative velocity becomes $W_1 = 1.225 U_1$, and the velocity head corresponding to the relative velocity can be related to the shaft speed N rpm and the flow rate Q gpm, or the specific speed N_s in English units and the head H ft by substituting the optimal diameter from the expression above:

$$\frac{W_1^2}{2g} = 1.5 \frac{U_1^2}{2g} = 2.12 \frac{U_1 C_1}{2g} = 1.153 \times 10^{-5} N^{4/3} Q^{2/3}$$
$$= 1.153 \times 10^5 N_s^{4/3} H$$

$$\text{specific speed} = N_s = \frac{N Q^{1/2}}{H^{3/4}} \quad (\text{rpm-gpm}^{1/2}/\text{ft}^{3/4})$$

The expression must have this particular form, to balance the dimensions, as can easily be verified. However, the coefficient of proportionality above implies the use of English units: rpm, gpm, and feet.

If pump specifications require a low NPSHR value, then minimizing the inlet relative velocity may not have first priority, and the possibility of cavitation must be examined in detail. Selection of the inlet diameter, blade angles, and location of the leading edge of the blades must proceed by trial and error. A trial design is assumed. The velocity along the leading edge can be calculated with the procedure described in Chapter 12 for inlet blade angle selection, which was programmed in the computer code INLET. Knowing the inlet geometry and inlet velocities, the NPSHR value can be calculated from the procedure described in Chapter 9. Of even greater importance than the NPSHR at design conditions might be the NPSHR off design, at the greatest flow rate the pump is likely to encounter. The flow will probably start to cavitate first at the shroud inlet. As will be seen, the manufacturing requirements imposed on the detailed blade design may require that the trial inlet geometry be changed. Cavitation calculations must then be repeated for the new geometry.

INLET BLADE ANGLE

As discussed above, the choice of the inlet diameter D_1 and the velocities at the inlet C_1 and U_1 also determine the inlet blade angle β_1. However, the inlet blade angle usually does not remain constant along the leading edge and must be calculated at several locations. In the case of a radial leading edge, the circumferential velocity decreases with decreasing radius along the leading edge in proportion to the radius, while the axial velocity remains the same. Therefore, the inlet blade angle β_1, measured from the axial direction, becomes smaller toward the axis of rotation of the impeller. The blades become steeper.

For a first, approximate design, an inlet blade angle β_1 of 55° from the meridional direction, or up to 65° on low-specific-speed pumps, should be

140 PUMP DESIGN PROCEDURE

adopted on the mid-streamline. The chosen axial inlet velocity usually conforms to the velocity in the standard inlet piping, which is typically sized for a velocity of $C_1 = 5$ to 25 ft/sec and is given in piping handbooks. Execution of the performance calculation computer program with a trial inlet angle will show if incidence losses have been minimized at the design flow rate. Artificial adjustments can be made in the computer program input. The actual, detailed inlet design should be the subject of a separate effort and must be combined with the detailed design of the impeller blades, as described below. After a first trial, a correction should also take into account the blockage due to the blade thickness at the inlet. If the length of the leading edge is s, and Z blades, t thick, are present, the inlet area in the calculation must be reduced by the amount of $sZt/(\cos\beta_1)$.

In practice, the leading edge of the blades is not radial but is swept back, as shown in Fig. 11.3. The flow becomes three-dimensional. Because of the curvature of the streamlines, the meridional inlet velocity becomes higher at the shroud than on the hub. Often, the leading edge is not perpendicular to the direction of the inlet velocity. However, the inlet blade angle must still be measured in the plane of the relative velocity W_1 and its meridional projection $C_1 = W_{m1}$. Obviously, the velocities and the blade angle will vary along the leading edge. The numerical value of the blade angle must be calculated separately at three locations at least: the hub, midspan, and shroud.

Impeller blade drawings usually show the projection of the blade on a radial (r–θ) plane, which is perpendicular to the axis of rotation. The blade angle on the drawing, β_{r1}, measured from the radial direction, differs from the actual blade

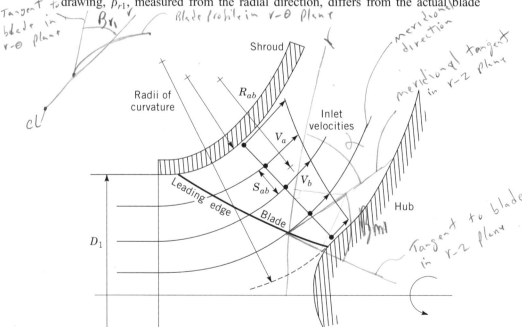

Figure 11.3 Impeller inlet velocity distribution.

angle, β_{m1}, measured from the meridional direction. The meridional direction corresponds to the tangent of the streamline in the axial–radial ($r-z$) plane, shown in the impeller cross section in Fig. 11.3. If the meridional direction makes an angle α with the radial direction, the actual blade angle is given by the relationship

$$\frac{U_1}{C_1} = \tan \beta_{m1} = \frac{\tan \beta_{r1}}{\cos \alpha}$$

Therefore, the actual blade angle, measured from the meridional direction, is larger than the blade angle in the projection.

An exact, axisymmetric potential flow calculation of the velocities and streamlines would show that the local pressure would increase and velocity would decrease perpendicular to the streamlines, away from the center of curvature of the streamlines, due to the centrifugal acceleration. The smaller the radius of curvature of the streamline, the greater will be the velocity change. High velocities will prevail at the shroud and low velocities at the hub. A sharp curvature at the shroud will induce particularly high velocities and should be avoided.

In the absence of a potential flow computer calculation, an approximate allowance can be made for the decrease of the velocity toward the hub, and increase toward the shroud, from an average velocity at the mid-streamline, which can be calculated from the flow rate and the flow passage cross section. If V_a and V_b are the velocities on two neighboring streamlines spaced at a distance of s_{ab} and having an average radius of curvature R_{ab}, the ratio of velocities is given approximately by

$$\frac{V_b}{V_a} = 1 - \frac{s_{ab}}{R_{ab}}$$

Subdividing the space between the hub and shroud contours by several streamlines, approximate values for the velocities can be calculated by applying stepwise the expression above from one streamline to the next (Fig. 11.3). The resulting velocities can be verified by calculating the flow rates between individual streamlines, or through these streamtubes. The sum of the individual flows should add up to the total flow rate. Such a simplified calculation procedure has been incorporated into the code INLET, presented in the Appendix.

The selection of the inlet blade angles and the location of the leading edge may have to be revised when the entire impeller blade is designed from inlet to outlet. Manufacturing considerations come into play, which may require some drastic compromises. However, blade angles in excess of 75° from meridional, or less than 15° from tangential, should be avoided. Such extreme angles would be encountered on the shroud. The relative velocities tend to become large, and the blade passage cross sections would become narrow. At extreme blade angles,

VANED DIFFUSER AND CROSSOVER INLET

Considerable kinetic energy remains in the flow leaving the impeller, a portion of which should be converted to pressure head with a diffuser (Fig. 11.4). In multistage pumps the flow leaving the impeller must be returned to the inlet of the next stage, and some pressure recovery is desirable in the process. The inlet section of the return passage often acts as a diffuser. The inlet leading edge of the diffuser, or of the return passage, must align with the direction of the absolute velocity leaving the impeller C_2. The exit velocity and its components in the radial or meridional and tangential direction are known from the impeller exit velocity diagram. The kinetic energy or head of the flow, and the desired diffuser vane angle β_3 measured from the meridional direction, are given by

$$\frac{C_2^2}{2g} = \frac{C_{r2}^2}{2g} + \frac{C_{t2}^2}{2g}$$

$$\tan \beta_3 = \frac{C_{t2}}{C_{r2}}$$

$$C_{r2} = \frac{Q}{\pi D_2 B_2}$$

$$C_{t2} = U_2 \sigma - C_{r2} \tan \beta_2$$

In these calculations the effective width B_2 should be assumed smaller than the actual geometrical measurement, to allow for blockage. If the diffuser closely

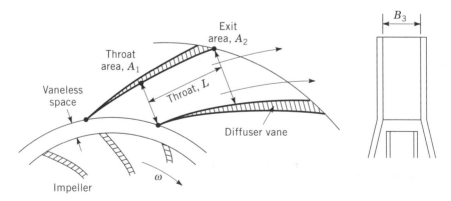

Figure 11.4 Vaned diffuser.

follows the diffuser [impeller], the diffuser inlet throat must accommodate the flow rate at the exit velocity C_2, which determines the diffuser width, or vane height B_3:

$$Q = C_2 \pi D_3 B_3 \cos \beta_3$$

$$B_3 = \frac{Q}{C_2 \pi D_3 \cos \beta_3}$$

Note that consistent units must be used in these expressions, and a conversion factor may be necessary, since velocities are usually given in ft/sec, the impeller geometry in inches, and the flow rate in gpm [$Q(\text{in}^3/\text{sec}) = Q(\text{gpm})(231\text{in}^3/\text{gall})(1\text{min}/60\text{sec})$].

The diffuser width B_3 can be corrected for vane blockage. If there are Z_3 vanes, t thick, the width becomes

$$B_3 = \frac{Q}{C_2(\pi D_3 \cos \beta_3 - Z_3 t)}$$

The inlet of the vaned diffuser follows closely the impeller exit, the radial distance being perhaps 5% of the impeller radius. In the case of a return passage, more space is left, and a correction needs to be made when calculating the inlet velocity, as explained below.

The number of diffuser or return passage vanes usually exceeds by one the number of impeller blades. If the number of blades were the same, all blades would pass the vanes at the same moment, the corresponding pressure fluctuations would be synchronized, and the blade passing noise and pressure fluctuations would be amplified. The intensity of the blade passing noise also depends on the spacing between the impeller and the diffuser inlet.

A radial vaned diffuser can consist of passages with straight, diverging walls, inclined to the radial direction as shown in Fig. 11.5. The diffuser area ratio usually remains below 2. As an alternative, the diffuser passages can be formed by airfoils, or slightly curved, circular partitions, arranged in a circle. The location immediately after the impeller exit, where the velocity is high, offers the greatest potential for pressure recovery. Even a modest diffusion can have a significant

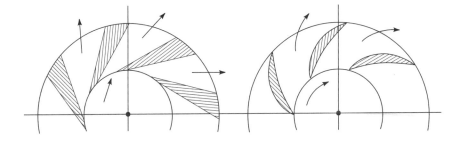

Figure 11.5 Vaned diffuser configurations.

effect. If the radial space is limited, widely spaced, thin, curved partitions or airfoils are used close to the impeller exit, to recover some velocity head.

Return passages in vertical column pumps are typically oriented in the axial direction, and the diameter of their centerline at the inlet D_3 is larger than the impeller diameter D_2. Therefore, the tangential component of the impeller exit velocity decreases in inverse proportion to the diameter, conforming to the principle of conservation of angular momentum. The return passage inlet velocity C_3 becomes

$$C_{t3} = \frac{C_{t2} D_2}{D_3}$$

$$C_3 = \left[C_{r2}^2 + \left(\frac{C_{t2} D_2}{D_3} \right)^2 \right]^{1/2}$$

$$\tan \beta_3 = \frac{C_{t2}}{C_{r2}} \frac{D_2}{D_3}$$

Often a wide space remains between the impeller exit and the return passage inlet. It can be assumed that the fluid rotates in this space with a constant angular momentum. This space merges with the clearance space between the hub and shroud faces of the impeller and the housing.

Since an imposed outer diameter often limits the radial space in multistage vertical pumps, the impeller exit can come close to the outer housing wall, and the flow must turn sharply to enter the return passage. The meridional velocity component is usually small; therefore, the turn does not result in significant losses. However, the proximity of the housing wall can restrict the impeller exit, which results in more slip than would be expected, and less head. On some radial pump designs the rear, hub face of the impeller exit is cut back to facilitate the turn into the return passage. The impeller exit then slopes backwards. The effective impeller diameter becomes somewhat smaller, but an optimal compromise may still be obtained. At higher specific speeds, mixed flow impellers are used with a slanted exit, in which the flow has a significant axial component.

VOLUTE DESIGN

The volute, as the diffuser, transforms some of the kinetic energy of the flow coming from the impeller into pressure energy by slowing down the flow velocity. The inefficiency of this diffusion process, in the components following the impeller, usually accounts for the major part of the pump losses. Therefore, the volute design must be an integral part of the pump design and should not be an afterthought.

VOLUTE DESIGN   145

The volute consists of a spiral-shaped passage surrounding the impeller exit, and collects the flow from around the impeller periphery (Fig. 11.6). The volute cross section, of circular or trapezoidal shape, increases gradually from the volute tongue to the volute throat, where the flow exits from the pump casing. The circle, centered on the axis of rotation and tangent to the volute tongue, is called the *base circle*. Its diameter is made about 10% greater than the impeller diameter. At flow rates below the design flow rate, some of the flow returns into the volute, passing between the impeller and the tongue, instead of exiting through the volute throat. Therefore, the space between the impeller and the tongue must not be made too small.

In some designs, radial space permitting, a radial vaneless space, with parallel walls, exists between the impeller exit and the volute. In this case the volute tongue is located at a distance more than 10% of the impeller radius. The width of the vaneless space is made slightly larger than the impeller exit width, and only a small clearance remains between the impeller rim and the vaneless space walls at the inlet. The purpose of the vaneless space is to recover some pressure by slowing the flow, due to an increase in the radius, as explained below.

Ideally, the volute cross section should accommodate the flow, which leaves the impeller without any velocity change. This consideration implies that the

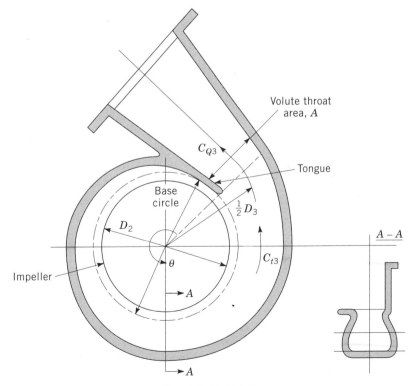

Figure 11.6 Volute.

cross-sectional area A should increase in proportion to the circumferential angle θ, measured from the tongue. However, since the center of the cross-sectional area moves out radially to increasingly larger radii, the circumferential, or tangential, velocity of the fluid will gradually decrease, in agreement with the principle of conserving its angular momentum. If the center of the volute throat, the final cross section above the tongue, is at a radius of $D_3/2$, the ideal volute crossection can be calculated from the expression

$$A = \frac{Q(\theta/360)}{C_{t\theta}}$$

where

$$C_{t\theta} = \frac{C_{t2}D_2}{D_2 + (D_3 - D_2)(\theta/360)}$$

Note that consistent units must be used since the flow rate is usually in gpm, the velocity in ft/sec, and the area in in^2.

In practice, the volute cross-sectional area A should be made greater by 15 to 25% than the ideal cross section, because losses slow the velocity and boundary layers build up. Also, it is easier to modify volute casting patterns to reduce than to increase the volute cross-sectional area. Volute design also depends on any diffuser and pressure recovery, which is expected to follow the volute throat. If the velocity in the volute throat has the same magnitude as the velocity at the pump exit flange, no further diffusion and pressure recovery can be expected. On the other hand, if an effective diffuser is to follow the volute, it makes sense to conserve kinetic energy and keep volute velocities relatively high in the volute. If the volute losses already slow down the velocity to its value at the exit flange, further diffusion, or attempt of pressure recovery, is futile. If there is no intention to recover the velocity head at the pump exit, it may as well be lost in the volute. The exit flange of standard process pumps is often located vertically on the centerline of the impeller, with a sharp bend between the volute throat and the exit flange. No diffusion and pressure recovery can be expected beyond the volute throat.

MECHANICAL DESIGN

The standards of the Hydraulic Institute, ISO, ASME (B73), and API (610) prescribe appropriate inlet and outlet pipe and flange sizes, overall housing configurations, sufficiently strong mounting arrangements, wear ring clearance sizes, and unbalance and vibration limits. Some of these standard measurements must be adopted directly, without questions, to satisfy particular customers. Other portions of the standards only prescribe the forces that the structure must withstand and leave it to the designer to select the configuration and to ascertain sufficient strength by stress calculations.

A general description of the mechanical design procedures, in particular stress calculations and rotor dynamic calculations, would far exceed the scope of this book and cannot be given here. However, this omission should not be construed to belittle their importance—on the contrary. Reliability and durability, which depend critically on sound mechanical design, have received great emphasis in recent years. They have become the most important concern of pump users, especially in the chemical and power generation industry, where service and maintenance costs over the life of the pump often exceed its first cost. A brief introduction to pump rotor dynamics is offered in Chapter 13.

Much research has been done on rotor dynamics at the Turbomachinery Laboratory of Texas A&M University or at ROMAC of the University of Virginia, for example (Barrett et al. 1978, Childs 1993, Corbo and Malanoski 1998, Vance 1988).

IMPELLER BLADE DESIGN

The desired suction head and correct incidence at the pump inlet determine the inlet blade angle β_1, and the desired pump head determines the exit blade angle β_2. The major dimensions of the impeller—inlet diameter D_1, exit diameter D_2, exit width B_2—have been fixed. These choices have been verified by executing the performance calculation computer program. The detailed contour of the hub and shroud and the blade angle distribution from inlet to exit remain to be designed.

The impeller blade angle distribution will have to be determined on at least three streamlines: the hub, shroud, and the mean streamlines shown in Fig. 11.7. In reality the flow rarely follows these paths but can twist around from the hub to the pressure or suction side of the blades, on to the shroud, or vice versa. Still, in a well-designed pump, the energy imparted to the fluid is about the same on all streamlines, and at least as an approximation, it can be assumed that the streamlines are interchangeable. Indeed, one of the guidelines of blade angle design is to impart the same energy on each streamline.

High-speed computer programs can calculate the flow velocities in great detail throughout the impeller. In particular, they can calculate the angular momentum rC_t, which is a measure of the energy added to the fluid at various locations along the blade. Ideal fluid mechanic theory predicts that the ideal blade angle distribution would be such that the angular momentum did not change perpendicular to the streamlines in the meridional $(r-z)$ plane (Tuzson 1993). The exact location of several streamlines and the velocities on them would have to be calculated with the computer. Such a calculation is often impractical, and the great effort would not be warranted in the case of most pumps. Instead, streamlines are assumed to follow the hub, shroud, and the mean path through the impeller. Sometimes, calculations are also made along the pressure and suction sides of the blades. The length of the streamlines are estimated, and the blades are designed to produce the same energy increase over the same length fraction of each streamline. Practical blade shapes require a compromise between the ideal

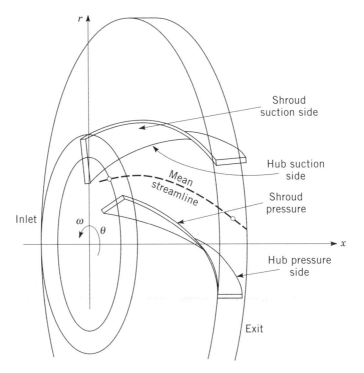

Figure 11.7 Impeller streamline locations for velocity and blade coordinate calculations.

conditions, the flow losses, and practical manufacturing considerations. The final design results from several iterations. After each iteration the design is checked for inlet incidence and cavitation, for appropriate exit conditions—in particular exit recirculation in mixed flow pumps—and above all, for ease and cost of manufacturing. The final compromise should conform to general industrial practices, or perhaps, even exceed these. Exceptional performance requires a combined analysis, design, and testing program, consisting of several iterations.

The best approach to determine the hub and shroud contours is to prepare a preliminary layout drawing. In radial impellers, when the inlet diameter D_1 is about half or less of the impeller diameter D_2, the shroud inlet radius of curvature should be about half of the inlet radius. As a first approximation one can assume that the meridional velocity, already estimated at the inlet and exit with the performance calculation program, declines slowly along the mean streamline. The needed approximate flow cross section and width B from hub to shroud, perpendicular to the meridional velocity vector, can be calculated from the meridional velocity W_r and the flow rate Q at a given radius r:

$$B = \frac{Q}{2\pi r W_r}$$

The leading edge will usually be slanted and should be drawn so that the average axial inlet velocity $C_1 = Q/(\pi D_1^2/4)$ should not change from the inlet flange to the leading edge. Once the location of the leading edge is chosen, approximate streamlines can be sketched in, and their radius of curvature can be estimated. This geometrical information can be used to calculate approximately the velocities at various locations along the leading edge, according to the procedure outlined above. The computer program INLET is provided for this purpose. A sharp curvature on the shroud will result in high velocities and possible early cavitation. The shroud curvature and leading-edge location can be revised to optimize inlet conditions. The meridional velocity C_{m1} and the circumferential velocity U_1 at various locations along the leading edge also define the blade angle β_1, which here is measured from the meridional direction, the direction of the velocity C_{m1}:

$$\tan \beta_1 = \frac{U_1}{C_{m1}}$$

It will be noted that along a slanted leading edge, the increase in the tangential velocity due to the increasing radius can compensate, at least partially, for the increase in meridional velocity due to the curvature of the streamlines. The meridional blade angle is not the angle shown on the usual view of the axial projection of the drawings, as explained above.

In the case of mixed flow pumps, when the flow path in the impeller is inclined toward the axis of rotation, the hub and shroud contours are often straight lines. There can be a curvature at the shroud inlet to provide a smooth transition. Ideally, the radius should be constant along the exit, the trailing edge of the blades. However, a slanted exit is often used, especially in mixed flow pumps. It can lead to exit recirculation, as discussed in Chapter 14.

MANUFACTURING REQUIREMENTS

Manufacturing methods profoundly affect impeller blade design. Prototype impellers are sometimes machined. With advanced, numerically controlled machine tools, just about any shape can be produced if cost is not an issue. Mass-produced impellers are most often cast. The metal casting and mold preparation process imposes severe limitations on the blade design. Often, it is not feasibility but cost which decides the use of a particular casting process. In the mold preparation process a wooden model is made of the impeller and its blades, casting sand is packed around the impeller model, and then the wooden model must be pulled from the mold, leaving a cavity that is an exact replica of the form to be cast. The geometrical shape of the blades must be such that it should be possible to pull them from the mold. Evidently, twisted blade shapes would be impossible to remove.

When the blade shape does not depend on the axial z coordinate, the simplest blade configurations are two-dimensional blades. The surface of all blades consists of axial generator lines. In the radial plane the blades are generally spiral shaped and are fully defined by the radial r and circumferential θ coordinates. The trace of the intersection of the blade with the hub and shroud surfaces overlap when viewed from the axial direction. Impeller models with such blades can be pulled from the mold in the axial direction, in one piece, with a simple movement which is easy to mechanize and relatively inexpensive to implement. Note that the leading edge need not necessarily be in the axial direction, nor need the hub and shroud contours be radial.

In a more complex mold configuration the blades are pulled out individually, in a direction inclined to the axial direction. This approach offers much greater flexibility for blade design but is more expensive to produce, requiring that each blade be handled separately. The surface of each blade consists of straight-line generators, all of which are inclined in the same direction, at an angle to the axial direction. The blades need to be slightly tapered in the direction in which they are pulled, which is usually toward the hub.

Finally, separate molds can be prepared for individual blades of quite arbitrary shape. The molds divide into two parts, one for each face of the blade, and the blades do not need to be pulled out. The individual blade molds can be assembled for casting. However, handling of the individual blade molds makes the process expensive. Also, the individual blade molds of the assembly can shift during casting, increasing the reject rate.

Exceptionally, investment (or lost wax) casting can be used when high accuracy is desired at any cost. A wax model of the impellers or blades is embedded in casting sand and the wax model is melted out, leaving a cavity for the metal. Manufacturing techniques more often used recently, such as stamping or polymer molding, impose special requirements which cannot be described in detail here. However, the design tools offered here are suitable for the design of such pumps—only the feasibility criteria differ.

BLADE COORDINATES

A list of coordinates—the radius r, axial distance z, and circumferential angle θ—on several streamlines determines the complex surface of the blade shape. These coordinates are defined on drawings by showing the intersection curves with the hub, the shroud, and also a curve along the mid-streamline. In impellers with two-dimensional blades, the hub and shroud lines coincide, the axial coordinate z is constant, and the other two coordinates suffice to define the blade shape. One of the variables, usually r or z, is selected as the independent variable. Its extent, from inlet to exit, is divided into 10 or 20 equal increments. Before the value of the other variables can be calculated at each station, the desired stepwise increases in the angular momentum rC_t along the streamlines need to be specified.

The incremental change in the angular momentum corresponds to the pressure difference exerted on the blade, the blade loading. Ideally, it should increase gradually from the inlet, reach a maximum at the middle of the blade length, and taper off at the exit (Tuzson 1993). Heavy blade loading at the inlet may increase inlet losses. Heavy blade loading at the exit may create exit losses, pressure fluctuations, and noise. Therefore, the angular momentum is gradually incremented from the inlet to its exit value, R_2C_{t2}. The local blade angle, measured from the radial direction or from the meridional direction in mixed flow impellers, can be calculated from the following geometrical relationship, illustrated in Fig. 11.8:

$$\tan \beta = \frac{r^2 \omega \sigma - rC_t}{rW_m}$$

where the angular momentum rC_t is a stepwise-increasing fraction of its value at the impeller exit:

$$R_2 C_{t2} = \frac{gH}{\eta \omega}$$

The slip σ appears only toward the exit, and gradually decreases from 1, at a distance from the periphery along the blade equal to the blade spacing to its value at the exit. If one were to assume an eddy relative to the impeller between the blades, which causes the slip, its effect would extend over a distance on the order of the blade spacing. The value of the meridional or radial relative velocity W_m

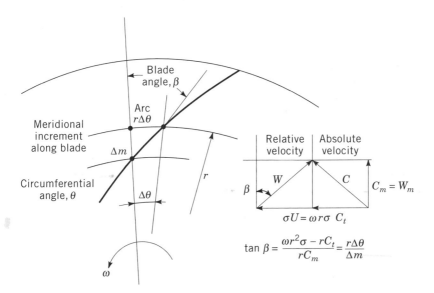

Figure 11.8 Local blade geometry and blade angle.

varies at the inlet from hub to shroud, and has been calculated when the inlet angles were determined. After the inlet bend, after the flow has turned radial—or approximately radial in mixed flow impellers—the relative meridional flow velocity can be assumed to be uniform across the flow passage and can be calculated from the flow rate and the passage cross-sectional area, corrected for blade and boundary layer blockage. It should be apparent that a computer program, calculating the meridional velocity in the impeller, would facilitate the design. However, since several iterations and adjustments are needed to arrive at a satisfactory blade shape, the informed judgment of the designer remains indispensable and may override theoretical considerations.

Having the blade angle β distribution along the blade, the increments of the variables Δr, Δz, and $\Delta \theta$ can be calculated from

$$\tan \beta = r \frac{\Delta \theta}{\Delta m}$$

where

$$\Delta m = (\Delta r^2 + \Delta z^2)^{1/2} \quad \text{or} \quad \Delta \theta = \frac{\Delta m \tan \beta}{r}$$

Note that the angle must be in radians: θ radians = θ degrees $(\pi/180)$. The tabulated blade coordinates are suitable for computer processing and can be transferred to plotting programs, which not only allow viewing of the blade shapes from different directions but also uncover any numerical error and unevenness of the shapes. These blade calculations are relatively simple but repetitive and are best executed with an appropriate computer program. Examples of such a programs are presented in Chapter 12. Since the calculations are simple, they can easily be checked by hand.

The resulting blade contour should change gradually from inlet to exit. It is not unusual to find that the inlet and exit blade angles are about the same, and consequently, a constant blade angle can be maintained along the blade. Such radial, constant-angle, logarithmic blades allow closed-form analytical flow calculations. The blade shape calculated in this manner may not be practical. The blade wrap, the cumulative circumferential angle from inlet θ_1 to outlet, θ_2, may be excessive. It should not exceed 100 to 120°. If the hub line is different from the shroud line, the blade wrap may also be different for each. The leading or trailing edge of the blade will then be inclined from the axial direction. Some slant can be tolerated, but excessive inclination produces small angles between the hub or shroud surfaces and the blade. Ideally, the blades should be approximately perpendicular to the hub and shroud surfaces.

Finally, a check is necessary to determine whether the blade can be pulled from the casting mold. With a computer plotting program, such a check can be performed by plotting both the blade curve on the hub and the curve on the shroud onto the same graph in three dimensions, and by turning the image until the two curves overlap. If the two curves cannot be made to overlap, it will be impossible to pull the blade from the mold. Some allowance can be made for the increasing blade thickness from shroud to hub, the casting draft, which will be applied. The casting technique dictates the blade thickness. Sufficient thickness must be provided for the passage of molten metal flow, which fills the cavity left by the withdrawn blade pattern. Often, especially for mixed flow pumps, the curves cannot be made to overlap. In such cases—or if the blade is not satisfactory for some other reason—a revision will be needed. A study of the first trial design will suggest possible modifications which can be tried in the next cycle of iteration. Typically, several iterations will be needed until a satisfactory blade shape is found. Not only the angular momentum distribution, but also the hub and shroud profile as well as the location of the leading edge, may have to be changed. High-specific-speed mixed flow impellers are considerably more difficult to design than radial impellers. Compromises are necessary, and extreme specifications may be unfeasible or too expensive, because complex casting procedures would be required.

COMPUTER CALCULATION OF BLADE COORDINATES

Simple, two-dimensional blades for radial impellers are often specified on drawings, in the view of axial projection, by successive arcs of circles, as shown in Fig. 11.9. The blade angle β, here measured from the radial direction, is specified at certain radii r. The radius of curvature R_c of the arc of circle joining the locations on the two radii r_a and r_b, where the arc of circle forms the angles $90 - \beta_a$ and $90 - \beta_b$ with the tangential direction, can be calculated from

$$R_c = \frac{r_a^2 - r_b^2}{r_a \sin(90 - \beta_a) - r_b \sin(90 - \beta_b)}$$

This relationship is derived by applying the cosine rule to the triangles formed by the radius, the radius of curvature, and the enclosed angle, which is $90 - \beta$.

Modern computer drafting programs allow easy construction of the blade curves and can define the centers of curvature, as well as the circumferential locations, θ, where the arcs of circles join. The program listing BLADE2 can be implemented and executed to calculate these variables. The computer programs IMPEL, MIXED, and BLADE2 presented in the Appendix can help design impellers and bladed return passages. The programs are not deterministic and do not lead directly to a practical blade configuration. Blade design requires a compromise, and the informed judgment of the designer determines, whether the criteria have been sufficiently well met. But before the criteria can be applied, the trial blade geometry must be fully developed. The value of the computer program resides in executing the detailed calculations fast and allowing several iterations and trial designs to be examined in rapid succession.

154 PUMP DESIGN PROCEDURE

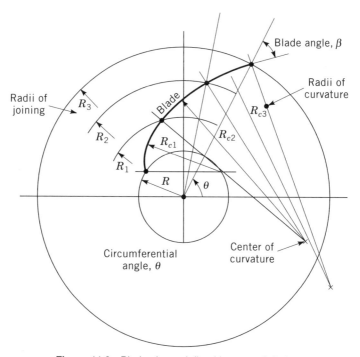

Figure 11.9 Blade shape defined by arcs of circles.

The IMPEL computer program is used to design blades that produce uniform angular momentum distribution, and therefore uniform energy addition, on all streamlines. The mean streamline coordinates in the meridional ($r-z$) plane—the flow rate, the inlet and exit meridional velocities, and the angular momentum at the exit—are specified in the input file. The program calculates the meridional flow passage width B and the hub and shroud r and z coordinates from the flow rate and the meridional velocity, which is assumed to change linearly, in equal increments, between the inlet and exit stations. The local value of the angular momentum, following a sinusoidal distribution from zero at the inlet to its specified value at the exit, determines the circumferential coordinates of the streamlines. The angular momentum being defined, division by the radius, gives the absolute tangential velocity, and subtracting the rotational speed, the relative tangential velocity is obtained. The tangent of the local blade angle β corresponds to the ratio of the relative tangential and the meridional velocities. Once the local blade angle is known, the circumferential, wrap angle ($\theta_2 - \theta_1$) can be calculated by adding up all its increments $\Delta\theta$, calculated from

$$\Delta\theta = \frac{\Delta m \tan \beta}{r} = \frac{\Delta m W_t / W_r}{r}$$

$$\Delta m^2 = \Delta r^2 + \Delta z^2$$

COMPUTER CALCULATION OF BLADE COORDINATES   155

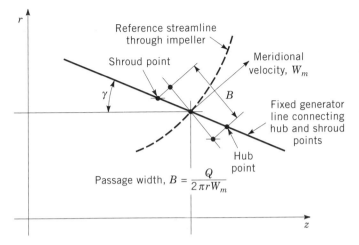

Figure 11.10 Hub and shroud coordinate calculation along a blade generator line in the meridional plane.

In the output file the blade coordinates, angular momentum, and blade angles on shroud, mean, and hub are printed out.

The MIXED program designs blade surfaces generated with parallel straight lines, inclined to the axial direction, and starting from a reference streamline. The generator lines and geometry are shown on projections in the $r-z$ and $r-\theta$ planes in Figs. 11.10 and 11.11, respectively. Parallel generator lines assure that the blade shape can be pulled from the casting mold in the direction of the generators. This direction can be chosen arbitrarily, but preferably should be approximately perpendicular to the hub and shroud surfaces near the middle of the blade. The angular momentum increase, the energy addition, is specified on the reference streamline. However, on the other streamlines the angular momentum distribution

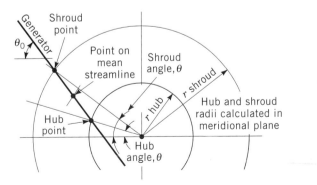

Figure 11.11 Hub and shroud coordinate calculation along a blade generator line in the circumferential plane.

156 PUMP DESIGN PROCEDURE

will generally differ. The blade angles along the hub and shroud streamlines are determined by the requirement of parallel generator lines, and a particular angular momentum distribution cannot be imposed. This is where the judgment and skill of the designer come into play, to ascertain that the energy addition on other streamlines, although different, remains acceptable.

The computer program listing MIXED in the Appendix contains the actual calculation procedure. As shown in Fig. 11.10, first a flow passage width B in the meridional $(r-z)$ plane is calculated, perpendicular to the meridional velocity W_r. Next the angle between the perpendicular to the mean streamline and the chosen direction of pull in the meridional plane are determined. The hub and shroud coordinate points in the meridional plane will lie in the chosen pull direction, at a distance from the mean streamline point, which corresponds to the half passage width $B/2$ divided by the cosine of the angle just calculated.

The coordinates in the $(r-z)$ plane being determined, the calculation of the circumferential angle remains. As shown in Fig. 11.11, this time the circumferential angles of the hub and shroud coordinate points, which lie along the projection of the chosen pull direction in the circumferential $(r-\theta)$ plane, need to be determined. The hub and shroud coordinate locations in this projection lie at the known hub and shroud radii along the line in the direction of pull passing through the coordinate points. Some trigonometry and application of the cosine rule implement the numerical calculation, which can be followed using the program listing in the Appendix.

The criterion of parallel generating lines is conservative. Skilled casting workers can gently twist and rotate the blade pattern while they are pulled from the mold and can get away with slight deviations from the ideal shape. The taper of the blade thickness, the casting draft, and a fillet at the hub surface can also help. The blade is usually pulled toward the hub. A casting trial will be made before production begins. Close cooperation between the designer and the casting shop can avoid the need for later modifications.

Design of the leading and trailing edges requires a further compromise. The directions of the leading and trailing edge of the blade usually do not coincide with the direction of the generators, as shown in Fig. 11.12. The computer output can be modified so that other streamlines on the blade surface, than the reference streamline, do not start or end at the first and last stations of the blade calculation. Coordinate stations can be added to the hub or shroud lines beyond the ones calculated by the program, and the number of stations need not be the same on all streamlines for the purposes of preparing engineering drawings. Hand adjustments are required. These allow certain blade angle distributions at the leading and trailing edges of the hub and shroud streamlines, which do not affect the other streamlines and which may help approximate the desired criteria. The computer program MIXED presents blade starting points, which are along the straight-line generators. The corresponding blade surface needs to be tailored and modified by hand to accommodate slanted leading or trailing edges. The inlet blade angles along the leading edge, calculated by the MIXED program, will not coincide with the values previously calculated, and the trailing edge blade angles

COMPUTER CALCULATION OF BLADE COORDINATES 157

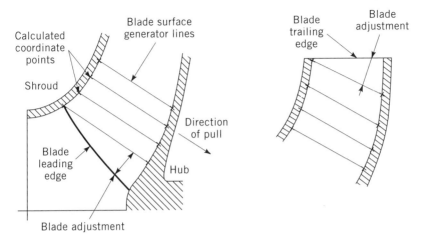

Figure 11.12 Blade adjustments at the leading and trailing edges.

may also differ. Adjustments can be made, for example, on the hub streamline if the effective length of the hub streamline, from leading to trailing edge, is longer than the effective length of the shroud streamline. The hub streamline will emerge from the mold before the rest of the blade, and the blade angles on both ends can be chosen without affecting the rest of the blade.

An approximate estimate of the meridional velocity W_{m1} distribution at the blade leading edge can be obtained by executing the INLET computer program. Ideally, complete, polar, axisymmetic partial differential equations should be solved, which, however, require that the hub and shroud contours be defined with very detailed smooth lists of coordinates. The partial differential equations operate with the higher derivatives of the variables. In numerical calculations the first derivatives are approximated by the finite differences of the variables, and the second derivatives, by the differences of these differences. Slight errors or inaccuracies from the finite difference approximation and from the rounding off of the numbers can propagate and amplify throughout the calculated flow field, leading to poor convergence of the iterations. For example, the radii of curvature of the streamlines, entering into the calculations, depend on the higher derivatives of the streamlines, the location of which themselves are unknown and need to be calculated.

The approximate calculation procedure incorporated into the code INLET assumes that a preliminary layout of the meridional cross section of the impeller has been prepared, and that, therefore, estimates of the radii of curvatures of the hub and shroud contours in the neighborhood of the inlet are available. The input file INLET.DAT needs to contain the coordinates of the hub and shroud streamlines, the coordinates of the leading edge on hub and shroud, the radii of curvature of the hub and shroud contours, and the flow rate. Equally spaced streamline locations are defined across the inlet cross section, and their radii of curvature are calculated from the hub and shroud radii of curvature by linear interpolation. The meridional velocity, from one streamline location to the next, increases by the factor (1+distance between streamlines/radius of curvature), as explained above. The program prints the calculation results, the values of the meridional velocity WM1 across the inlet cross section, into the same file INLET.DAT. The possibility of cavitation at the inlet must be reconsidered in the light of higher shroud velocities, and a new value of NPSHR should be calculated. Cavitation appears first at the shroud inlet.

The computer program MIXED can also be used to design return passages with stationary vanes for multistage vertical pumps. Obviously, there is no rotation $\omega = 0$, and the absolute tangential velocity C_t must gradually decrease from its value at the inlet to the outlet. The vanes at the exit must point in the axial direction to eliminate any swirl that may enter the next stage. The meridional contour takes on an "S" shape, from axial at the inlet to axial again at the exit, near the shaft. In the case of low-specific-speed pumps, when the flow passages are relatively narrow, the return passage can be made radial and simple two-dimensional vanes become possible. The inlet leading edge is sometimes at an extreme slant, reaching over the impeller exit, along the outer housing wall.

The blade shapes must somehow be communicated to the patternmaker. In computer-based drafting, the blade coordinates, r, z, and θ are specified on the hub and shroud contour. The coordinates are listed on the impeller drawings but should preferably be transferred to the patternmaker by the computer disk, because data input labor can then be minimized and numerical errors avoided. Older practice, still often used, requires that several blade cross sections be shown on the impeller drawings. The cross sections are taken in planes defined by a constant axial coordinate, $z =$ constant, the location of successive cuts being at axial coordinates z incremented by equal amounts. The intersection of such a plane with the blade surface is defined by the curve relating the radius r to the circumferential angle θ. The patternmaker builds up the blade model from layers of parallel equal-thickness boards, which are cut to the required profile, and stacked. The assembly is then sanded smooth.

The computer program CUT presented in the Appendix executes such cuts. The blade shapes are defined by the intersections of the generator lines and the hub and shroud surfaces, which are the hub and shroud coordinates given in the input file. The first step in the calculation procedure consists of determining the r coordinates corresponding to the chosen z coordinates in the meridional $r-z$ plane. The r coordinate is given by the intersection of the z plane and the projection of the generator line between the hub and shroud points. In the circumferential $r-\theta$ plane the circumferential θ coordinate is given by the intersection of the generator line projection between hub and shroud points and the circle of the previously calculated radius r. The actual angle is calculated using the sine rule. Calculation details are contained in the program listing.

When the blade design has been completed satisfactorily, the entire design process should be repeated, and key variables should be checked by hand calculations to ascertain that the simplifications that have been made in the design process have not distorted the geometry, that no mistakes were made, and that the design will indeed perform satisfactorily.

REFERENCES

Barrett, L. E., Gunter, E. J., Allaire, P. E. (1978): Optimum Bearing and Support Damping for Unbalance Response and Stability of Turbomachinery, *ASME Journal of Engineering for Power*, January, pp. 89–94.

Childs, D. (1993): *Turbomachinery Rotordynamics*, Wiley, New York.

Church, G. (1944): *Centrifugal Pumps and Blowers*, Wiley, New York.

Corbo, M. A., Malanoski, S. B. (1998): Pump Rotordynamics Made Simple, *Proceedings of the 15th International Pump User Symposium*, Texas A&M University, College Station, Texas, pp. 167–204.

Hydraulic Institute (1994): *Centrifugal Pump Standards*, HI, Parsippary, N.J.

Oak Ridge (1972): *Design and Procurement Guide for Primary Coolant Pumps in Light-Water-Cooled Nuclear Reactors*, Government Report ORNL-TM-3956, November.

Peck, J. F. (1968): Design of Centrifugal Pumps with Computer Aid, *Proceedings of the Institution of Mechanical Engineers*, Vol. 183, Pt. 1, No. 17, pp. 321–351.

Pfleiderer, C. (1961): *Die Kreiselpumpen*, Springer-Verlag, Berlin.

Rey, R., Kermarec, Y., Guiton, P., Vullioud, G. (1982): Étude statistique sur les caracteristiques à debit partiel des pompes centrifuges et sur la determination approximative du debit critique (Statistical Study of Partial Flow Characteristics of Centrifugal Pumps and of Approximate Determination of the Critical Flow), *La Houille Blanche*, No. 2/3, pp. 107–120.

Salisbury, A. G. (1983): Current Concepts in Centrifugal Pump Hydraulic Design, *Proceedings of the Institution of Mechanical Engineers*, Vol. 197A, October, pp. 221–231.

Spring, H. (1983): A Comprehensive One-Dimensional Streamline Theory Based on Euler, with Pfleiderer's Modification for Slip and Hydraulic Efficiency, Suitable for Centrifugal Pumps with Prerotation, in *Performance Characteristics of Turbines and Pumps*, ASME FED Vol. 6, pp. 141–147.

Spring, H. (1984): *The Impeller Inlet Diameter Method to Justify Proper Selection for Pumps*, ASME Paper 84-WA/FM-1.

Turton, R. K. (1994): *Rotordynamic Pump Design*, Cambridge University Press, New York.

Tuzson, J. (1993): Interpretation of Impeller Flow Calculations, *ASME Journal of Fluids Engineering*, September, pp. 463–467.

Vance, J. M. (1988): *Rotordynamics of Turbomachinery*, Wiley, New York.

Vlaming, D. J. (1989): Optimum Impeller Inlet Geometry for Minimum NPSH Requirement for Centrifugal Pumps, in *Pumping Machinery*, ASME FED Vol. 81, pp. 25–28.

12

DESIGN EXAMPLES

PRELIMINARY DESIGN AND PERFORMANCE ESTIMATE

The objective of this chapter is to illustrate the design process with numerical examples. Since a first design trial is verified with the help of the LOSS3 performance prediction computer program, and since the computer programs in the Appendix are used to calculate the detailed blade geometry, in many instances the symbols of the FORTRAN listings are also used in the text, with the expectation that this approach will facilitate interpretation of the computer printouts. In the text the symbols are preceded by their verbal description. A full nomenclature can be found in the Appendix, and often in the listings themselves.

As an example, the design of a pump producing 100 ft (30.5 m) head at 2000 gpm (0.126 m^3/s) when driven at 1770 rpm is considered. It is a vertical type with a mixed flow impeller and combined diffuser and return passage. The foregoing specifications correspond to a specific speed of 2500 in English units. The efficiency of similar commercially available pumps would be 86% based on the specific speed and flow rate. Assuming a head coefficient ψ on the order of 0.5, the relationship between head H_0 and tip speed shows that a tip speed of about 86 ft/sec (26.32 m/s) might be appropriate. At 1770 rpm such a tip speed would be reached with an 11-in.-diameter impeller.

$$H_{th} = \frac{H_0}{\eta} = \frac{U_2^2 \psi}{g}$$

$$U_2 = \left(\frac{H_0 g}{\eta \psi}\right)^{1/2} = \frac{\omega D_2}{2}$$

$$U_2 = \left[\frac{100 \text{ ft}(32.2 \text{ ft/sec}^2)}{(0.86)(0.5)}\right]^{1/2} = 86.5 \text{ ft/sec}$$

$$U_2 = \left[\frac{30.5 \text{ m}(9.8 \text{ m/s}^2)}{(0.86)(0.5)}\right]^{1/2} = 26.32 \text{ m}$$

$$\omega = [2\pi/(60 \text{ sec/min})]1770 \text{ rpm} = 185.35 \text{ sec}^{-1}$$

$$D_2 = \frac{2(86.5 \text{ ft/sec})(12 \text{ in./ft})}{[2\pi/(60 \text{ sec/min})]1770 \text{ rpm}} = 11.2 \text{ in.}$$

$$D_2 = \frac{2(26.32 \text{ m/s})}{[2\pi/(60 \text{ sec/min})]1770 \text{ rpm}} = 0.284 \text{ m}$$

Assuming a flow coefficient ϕ of 0.15, an impeller width B_2 of 1.4 in. would result in a flow rate of 2000 gpm.

$$Q = (\pi D_2 B_2) U_2 \phi$$

$$B_2 = \frac{Q}{\pi D_2 U_2 \phi}$$

$$B_2 = \frac{2000 \text{ gpm}(231 \text{ in}^3/\text{gal})}{(60 \text{ min/sec})\pi(11.2 \text{ in.})(86.5 \text{ ft/sec})(12 \text{ in./ft})0.15} = 1.406 \text{ in.}$$

$$B_2 = \frac{0.126 \text{ m}^3/\text{s}}{[\pi(0.284 \text{ m})(26.32 \text{ m/s})0.15]} = 0.0358 \text{ m} = 3.58 \text{ cm}$$

As a start, the optimum inlet diameter, corresponding to an inlet blade angle β_1 of 54° from meridional, should be calculated from

$$D_1 = 1.533 \left(\frac{Q}{\omega}\right)^{1/3} = 1.533 \left(\frac{7700 \text{ in.}^3/\text{sec}}{185.35 \text{ /sec}}\right)^{1/3} = 5.3 \text{ in.}$$

$$Q = 2000 \text{ gpm}(231 \text{ in.}^3/\text{gal})/(60 \text{ sec/min}) = 7700 \text{ in.}^3/\text{sec}$$

$$D_1 = 1.533 \left[\frac{0.126 \text{ m}^3/\text{s}}{2\pi 1770 \text{ rpm}/(60 \text{ sec/min})}\right]^{1/3} = 0.135 \text{ m}$$

The exit blade angle turns out to be about 65° from meridional or 25° from tangential, whether read from the head–flow coefficient plot or—assuming a slip coefficient of 0.82—calculated from

$$\tan \beta_2 = \frac{\sigma - \varphi}{\phi} = \frac{0.82 - 0.5}{0.15} = 2.133$$

The return passage average diameter D_3 should be 12.75 in. (0.324 m) which is 14% more than the impeller diameter, and its vane height B_3 should be 1.0 in. (2.54 cm), somewhat smaller than the impeller exit width. In these calculations the flow passage widths correspond to the effective width available to the flow. The geometrical width may have to be somewhat larger to allow for blade and boundary layer blockage. These measurements must be checked against the available space on a layout. Diameter limitations may be imposed by the maximum permissible outer diameter of the casing. As a first try, the diffuser inlet angle can be taken equal to the angle of the flow leaving the impeller, which is given by

$$\tan \beta_3 = \frac{\varphi}{\phi} = \frac{0.5}{0.15} = 3.333$$

$$\beta_3 = 73.3°$$

To complete the input values for the performance calculation computer program LOSS3, six impeller blades and seven diffuser vanes are assumed. The loss coefficients should match the loss coefficients of a similar pump on which the program should have been calibrated previously.

The printout of the performance calculation computer program LOSS3 in Fig. 12.1, with the foregoing values of the input variables, shows the following:

1. The maximum efficiency, only 81.21%, is at 1500 gpm (0.0946 m³/s), where the diffuser approach velocity C3 = 43.01 ft/sec (13.14 m/s) and the diffuser throat velocity CQ3 = 41.81 ft/sec (12.74 m/s) are reasonably well matched. The diffuser throat velocity at 2000 gpm (0.126 m³/s) is excessive: CQ3 = 55.75 ft/sec (17.0 m/s), compared with the approach velocity C3 = 38.64 ft/sec (11.78 m/s). The high diffuser throat velocity also accounts for the high diffuser loss DQVD = 15.15 ft (4.6 m). Therefore, the diffuser throat needs to be opened up, which is accomplished by decreasing the inlet angle β_3 =BETA3 from 73.3° to about 70.0°, and increasing the diffuser width B3 from 1.0 to 1.2 in. (2.54 to 3.05 cm).
2. The head is low at 2000 gpm: DH = 83.22 ft (25.36 m). Therefore, the impeller diameter D2 could be increased from 11.00 in. to 11.60 in. (0.28 to 0.295 m), but the exit blade angle β_2 = BETA2 is increased to 68°, and the impeller exit width B2 is changed to 1.300 in. (3.3 cm).
3. The inlet incidence is perfect, the loss DQIN12 being 0.01 ft at 2000 gpm, the design point. However, the inlet absolute velocity C1 = 29.08 ft/sec

164 DESIGN EXAMPLES

PUMP PERFORMANCE - John Tuzson & Associates.1995.

EXAMPLE 1st TRY 1/18/99

```
Q       DN      D1      DS      D2      D3      B       B3      D4      AREA
2000.0  1770.0  5.300   0.000   11.000  12.750  1.400   1.000   0.000   0.000
BETA1   BETA2   BETA3   NB      NV      DISK    SKIN    RECIRC  CVD     CIN
54.00   65.00   73.30   6.0     7.0     0.0040  0.0045  0.0040  0.2000  0.4000

            Q(I)    DH      EFF             DHTH    DQIN12  DQSF12  DQDIF
            500.00  124.09  46.50           163.81  14.85   0.65    6.23
DQIN23  DQSF23  DQSF34  DQVD    DFH     DRECH           QL      U1      U2
16.71   0.00    0.33    0.95    7.17    87.01           17.24   40.93   84.95
C1      W1      WT1     W2      WM2     CT2     CT3     C3      CQ3     C4
7.27    41.57   10.01   7.85    3.32    62.09   53.57   53.71   13.94   0.00

            Q(I)    DH      EFF             DHTH    DQIN12  DQSF12  DQDIF
            1000.00 117.27  70.22           145.05  6.79    0.93    5.41
DQIN23  DQSF23  DQSF34  DQVD    DFH     DRECH           QL      U1      U2
9.55    0.00    1.31    3.79    3.59    15.79           15.67   40.93   84.95
C1      W1      WT1     W2      WM2     CT2     CT3     C3      CQ3     C4
14.54   43.44   20.02   15.69   6.63    54.98   47.43   48.10   27.87   0.00

            Q(I)    DH      EFF             DHTH    DQIN12  DQSF12  DQDIF
            1500.00 106.99  81.21           126.29  1.85    1.30    4.05
DQIN23  DQSF23  DQSF34  DQVD    DFH     DRECH   ·       QL      U1      U2
0.63    0.00    2.95    8.52    2.39    1.86            13.92   40.93   84.95
C1      W1      WT1     W2      WM2     CT2     CT3     C3      CQ3     C4
21.81   46.38   30.02   23.54   9.95    47.87   41.30   43.01   41.81   0.00

            Q(I)    DH      EFF             DHTH    DQIN12  DQSF12  DQDIF
            2000.00 83.22   75.67           107.53  0.01    1.76    2.14
DQIN23  DQSF23  DQSF34  DQVD    DFH     DRECH           QL      U1      U2
0.00    0.00    5.24    15.15   1.79    0.00            11.92   40.93   84.95
C1      W1      WT1     W2      WM2     CT2     CT3     C3      CQ3     C4
29.08   50.21   40.03   31.38   13.26   40.76   35.16   38.64   55.75   0.00

            Q(I)    DH      EFF             DHTH    DQIN12  DQSF12  DQDIF
            2500.00 53.28   58.84           88.77   1.29    2.34    0.00
DQIN23  DQSF23  DQSF34  DQVD    DFH     DRECH           QL      U1      U2
0.00    0.00    8.19    23.67   1.43    0.00            9.50    40.93   84.95
C1      W1      WT1     W2      WM2     CT2     CT3     C3      CQ3     C4
36.36   54.75   50.04   39.23   16.58   33.64   29.03   35.26   69.68   0.00

            Q(I)    DH      EFF             DHTH    DQIN12  DQSF12  DQDIF
            3000.00 15.43   21.62           70.01   5.67    3.03    0.00
DQIN23  DQSF23  DQSF34  DQVD    DFH     DRECH           QL      U1      U2
0.00    0.00    11.79   34.08   1.20    0.00            6.21    40.93   84.95
C1      W1      WT1     W2      WM2     CT2     CT3     C3      CQ3     C4
43.63   59.82   60.05   47.07   19.89   26.53   22.89   33.19   83.62   0.00
```

Figure 12.1 Printout of the performance calculation program LOSS3 for the first iteration for a pump design example.

(8.86 m/s) is high and could cause cavitation problems, if that were an issue. According to the expression given for design conditions, the NPSHR value would be

$$\text{NPSHR} = C_1^2/2\,g = \frac{0.5(29.08 \text{ ft/sec})^2}{(32.2 \text{ ft/sec}^2)} = 13.13 \text{ ft}$$

$$\text{NPSHR} = C_1^2/2g = \frac{0.5(8.86 \text{ m/s})^2}{(9.8 \text{ m/s}^2)} = 4.0 \text{ m}$$

PRELIMINARY DESIGN AND PERFORMANCE ESTIMATE

Correct incidence will still be maintained if the inlet diameter is made $D1 = 6.5$ in. (0.165 m), and the inlet blade angle $\beta_1 = \text{BETA1} = 68.0°$.

The printout of the second iteration, in Fig. 12.2, with corrected input values, shows that the maximum efficiency, which increased to 86.24%, is now at 2000 gpm (0.126 m^3/s) where the diffuser velocities C3 and CQ3 are equal. The head is 99.72 ft (30.4 m), and the absolute inlet velocity C1 dropped to 19.34 ft/sec (5.89 m/s), from which an NPSHR value of 5.8 ft (1.77 m) could be calculated. The performance calculation code LOSS3 does not estimate NPSHR. These values meet the design specifications.

```
PUMP PERFORMANCE - John Tuzson & Associates.1995.

EXAMPLE 2nd TRY 1/18/99

Q        DN      D1     DS     D2      D3     B      B3      D4     AREA
2000.0  1770.0  6.500  0.000  11.600  12.750  1.300  1.200   0.000  0.000
BETA1   BETA2   BETA3   NB     NV     DISK    SKIN   RECIRC  CVD    CIN
68.00   68.00   70.00   6.0    7.0    0.0040  0.0045 0.0040  0.2000 0.4000

        Q(I)    DH      EFF            DHTH    DQIN12  DQSF12 DQDIF
        500.00  127.30  37.77          182.41  22.70   1.02   9.24
DQIN23  DQSF23  DQSF34  DQVD   DFH     DRECH           QL     U1     U2
21.58   0.00    0.09    0.48   9.35    130.87          22.32  50.20  89.59
C1      W1      WT1     W2     WM2     CT2     CT3     C3     CQ3    C4
4.83    50.43   11.97   9.04   3.39    65.56   59.65   59.74  9.76   0.00

        Q(I)    DH      EFF            DHTH    DQIN12  DQSF12 DQDIF
        1000.00 122.37  63.98          159.10  10.72   1.38   7.61
DQIN23  DQSF23  DQSF34  DQVD   DFH     DRECH           QL     U1     U2
14.72   0.00    0.37    1.94   4.68    23.74           20.04  50.20  89.59
C1      W1      WT1     W2     WM2     CT2     CT3     C3     CQ3    C4
9.67    51.12   23.93   18.08  6.77    57.18   52.03   52.45  19.52  0.00

        Q(I)    DH      EFF            DHTH    DQIN12  DQSF12 DQDIF
        1500.00 113.18  78.96          135.78  3.18    1.81   4.89
DQIN23  DQSF23  DQSF34  DQVD   DFH     DRECH           QL     U1     U2
7.54    0.00    0.82    4.35   3.12    2.80            17.46  50.20  89.59
C1      W1      WT1     W2     WM2     CT2     CT3     C3     CQ3    C4
14.50   52.25   35.90   27.12  10.16   48.80   44.40   45.52  29.27  0.00

        Q(I)    DH      EFF            DHTH    DQIN12  DQSF12 DQDIF
        2000.00 99.72   86.24          112.46  0.08    2.33   1.08
DQIN23  DQSF23  DQSF34  DQVD   DFH     DRECH           QL     U1     U2
0.04    0.00    1.46    7.74   2.34    0.00            14.42  50.20  89.59
C1      W1      WT1     W2     WM2     CT2     CT3     C3     CQ3    C4
19.34   53.80   47.86   36.16  13.54   40.42   36.78   39.12  39.03  0.00

        Q(I)    DH      EFF            DHTH    DQIN12  DQSF12 DQDIF
        2500.00 70.40   77.02          89.14   1.44    2.93   0.00
DQIN23  DQSF23  DQSF34  DQVD   DFH     DRECH           QL     U1     U2
0.00    0.00    2.28    12.10  1.87    0.00            10.55  50.20  89.59
C1      W1      WT1     W2     WM2     CT2     CT3     C3     CQ3    C4
24.17   55.72   59.83   45.20  16.93   32.04   29.15   33.59  48.79  0.00

        Q(I)    DH      EFF            DHTH    DQIN12  DQSF12 DQDIF
        3000.00 34.26   50.77          65.83   7.24    3.62   0.00
DQIN23  DQSF23  DQSF34  DQVD   DFH     DRECH           QL     U1     U2
0.00    0.00    3.29    17.42  1.56    0.00            3.82   50.20  89.59
C1      W1      WT1     W2     WM2     CT2     CT3     C3     CQ3    C4
29.01   57.98   71.79   54.23  20.32   23.66   21.53   29.40  58.55  0.00
```

Figure 12.2 Printout of the performance calculation program LOSS3 for the second iteration for a pump design example.

Additional runs, with slight modifications, can ascertain whether optimal conditions have been reached. It must be kept in mind that the absolute value of the quantities calculated by the LOSS3 code can differ by 1 or 2% from actual values measured in tests, even if the loss coefficients have been checked against the test data from a similar pump. The program does not make allowance for boundary layer and blade thickness blockage, which would require that the actual geometrical measurements of the flow passages be somewhat larger than the calculated effective cross-sectional areas. Still, trial runs can show in which direction changes in input variables affect the calculated values, even if the absolute values are in doubt. Evidently, other, perhaps more detailed performance prediction codes could also be used instead of the LOSS3 code, but usually they will require a greater number of input variables, which would need to be chosen before the program can be executed.

INLET VELOCITY CALCULATION

A simple estimate of the meridional velocity WM1 distribution across the inlet cross section can be obtained by executing the INLET computer program. The input file INLET.DAT contains coordinates r and z of the hub and shroud contours, their radii of curvature near the inlet, the coordinates of the leading edge on hub and shroud, and the flow rate. In this example a radial impeller configuration was assumed.

The calculation example gives the radii of curvature RHOH $= 4.75 = 12$ cm, and RHOS $= 1.0 = 2.54$ cm, and a flow rate $Q = 1000$ gpm (0.063 m^3/s). The calculated meridional velocities across the inlet cross section, printed in file INLET.DAT and shown in Fig. 12.3, range from 6.374 ft/sec (1.942 m/s) on the

```
File INLET.DAT input and output to the INLET program.

 1.14,1.415,1.708,2.018,2.344,2.682,3.032,3.391,3.756,4.127,
 4.5,4.65,4.8,4.95,5.1,5.25,5.4,5.55,5.7,5.85
 3.359,3.612,3.843,4.05,4.232,4.388,4.518,4.619,4.692,4.735,
 4.75,4.75,4.75,4.75,4.75,4.75,4.75,4.75,4.75,4.75
 3.0,3.0123,3.049,3.109,3.191,3.293,3.412,3.546,3.691,3.846,
 4.0,4.2,4.4,4.6,4.8,5.0,5.2,5.4,5.6,5.8
 2.0,2.1564,2.309,2.454,2.588,2.707,2.809,2.891,2.951,2.988,
 3.0,3.0,3.0,3.0,3.0,3.0,3.0,3.0,3.0,3.0
 1.708,3.843,3.0123,2.156,4.75,1.0,1000

 John Tuzson and Associates, Evanston,IL 1/28/99

 Calcualted meridional Velocity Distribution from Hub to Shroud
 6.374   6.780   7.272   7.887   8.683   9.771   11.386  14.151  20.620
  RRL     ZRL     RTL     ZTL    RHOH    RHOS    Q gpm
 1.708   3.843   3.012   2.156   4.750   1.000   1000.0
```

Figure 12.3 Input and output file of the INLET computer program, calculating the inlet velocity distribution for a pump impeller design example.

hub to 20.620 ft/sec (6.285 m/s) on the shroud. Evidently, the high local velocity on the shroud will have to be taken into account in designing the blades if correct incidence at design conditions is to be achieved. The shroud velocity will also affect the appearance of cavitation and the NPSHR value of the pump. New values of NPSHR, with updated shroud velocity estimates, should be prepared. Cavitation will appear first at the shroud inlet.

The designer must exercise good judgment as to whether to accept very high shroud velocities. In the case of a very small shroud radius of curvature, especially if the flow decelerates rapidly on the shroud beyond the inlet, flow separation from the shroud contour should be suspected. Since the relative flow velocity also has a strong tangential component, separation can take the form of a leading-edge vortex, especially at off-design conditions at incidence. A very complex separated region and embedded trailing vortex may appear along the shroud and blade corner which defy accurate analysis and mathematical description. Preferably, strong curvature on the shroud should be avoided.

The average of these calculated meridional velocities, 9.93 ft/sec (3.027 m/s), compares with a value of 10.2 ft/sec (3.109 m/s) calculated from the flow rate and the inlet cross-sectional area as if the velocity were uniform. The close agreement shows that in this case the flow passage cross sections can be estimated sufficiently accurately from an average meridional velocity and the flow rate, even when the streamlines are curved.

DETAILED THREE-DIMENSIONAL BLADE DESIGN

Since the values of the preliminary variables have been determined, detailed design of the impeller blades can begin. The blade coordinate calculation code IMPEL requires the input values of 21 coordinate points of the mean streamline in the meridional $(r-z)$ plane through the impeller. The flow rate, rotational speed, inlet velocity, meridional exit velocity, absolute tangential velocity, and slip are available from the printout of the performance calculation program LOSS3. In addition, the blade thickness T needs to be specified. The blade thickness only enters into the calculation of a blockage when the meridional velocity is estimated from the geometrical cross section of the flow passage. Blade thickness will be on the order of 0.125 to 0.250 in. (3.175 to 6.35 mm) and should be selected in consultation with the casting manufacturer, since it has to allow a satisfactory flow of the molten metal.

As expected, the printout of the IMPEL program (Fig. 12.4) shows that indeed the angular momentum distribution on hub, mean, and shroud is the same. The wrap angle from inlet to exit varies from 82° on the hub to 100° on the shroud. A three-dimensional plot of the blade coordinates would, however, show that hub and shroud lines cannot be made to overlap, regardless of the direction from which they are viewed. Therefore, the blades could not be pulled from a mold, and individual blade molds would be the only practical casting method. The

168 DESIGN EXAMPLES

John Tuzson & Associates, Evanston, IL 2/1/99

Mixed flow impeller blade coordinates.
Same energy addition on all steamlines.

RT	ZT	THET	RM	ZM	THEM	RR	ZR	THER
3.575	0.932	0.000	3.050	1.870	0.000	2.525	2.808	0.000
3.696	1.039	7.144	3.187	1.947	6.959	2.679	2.855	7.100
3.817	1.146	14.200	3.325	2.024	13.799	2.833	2.902	14.004
3.938	1.251	21.093	3.463	2.101	20.436	2.987	2.951	20.605
4.061	1.355	27.765	3.600	2.178	26.807	3.139	3.001	26.831
4.184	1.457	34.172	3.737	2.255	32.869	3.291	3.053	32.639
4.309	1.557	40.288	3.875	2.332	38.596	3.441	3.107	38.012
4.435	1.655	46.099	4.012	2.409	43.977	3.590	3.163	42.948
4.562	1.751	51.604	4.150	2.486	49.014	3.738	3.221	47.463
4.690	1.844	56.812	4.287	2.563	53.720	3.885	3.282	51.583
4.819	1.936	61.741	4.425	2.640	58.120	4.031	3.344	55.344
4.950	2.025	66.420	4.562	2.717	62.244	4.175	3.409	58.791
5.082	2.112	70.883	4.700	2.794	66.134	4.318	3.476	61.973
5.215	2.197	75.169	4.838	2.871	69.834	4.460	3.545	64.947
5.349	2.280	79.325	4.975	2.948	73.395	4.601	3.616	67.770
5.484	2.362	83.399	5.113	3.025	76.870	4.741	3.688	70.505
5.620	2.441	87.441	5.250	3.102	80.316	4.880	3.763	73.213
5.757	2.518	91.506	5.387	3.179	83.789	5.018	3.840	75.959
5.896	2.594	95.646	5.525	3.256	87.347	5.154	3.918	78.806
6.035	2.668	99.916	5.662	3.333	91.048	5.290	3.998	81.815

Angular Momentum			Blade Angles		
0.008	0.033	0.054	70.698	67.852	64.074
0.334	0.353	0.365	71.069	68.346	64.726
0.873	0.883	0.884	71.240	68.548	64.936
1.614	1.613	1.601	71.243	68.511	64.793
2.538	2.525	2.499	71.103	68.271	64.354
3.622	3.599	3.560	70.844	67.860	63.657
4.840	4.807	4.758	70.486	67.305	62.738
6.163	6.122	6.063	70.054	66.637	61.635
7.558	7.510	7.445	69.577	65.892	60.396
8.990	8.938	8.869	69.087	65.117	59.085
10.425	10.371	10.299	68.625	64.369	57.790
11.827	11.772	11.701	68.236	63.715	56.621
13.161	13.108	13.039	67.970	63.232	55.705
14.394	14.344	14.280	67.872	62.992	55.170
15.496	15.451	15.393	67.982	63.059	55.131
16.439	16.400	16.349	68.323	63.473	55.657
17.199	17.167	17.125	68.901	64.240	56.763
17.758	17.734	17.700	69.700	65.332	58.391
18.101	18.085	18.061	70.683	66.688	60.430

Figure 12.4 Output file of the computer program IMPEL, calculating blade coordinates for a pump impeller design example, having the same angular velocity distribution on all streamlines.

individual blade molds would then have to be assembled and joined to form the impeller mold.

If the blades are to be pulled individually from an impeller casting mold, the computer program MIXED would have to be used. The direction of the pull, which will also define the straight-line generators, connecting corresponding points on hub and shroud, must then be chosen. In the circumferential direction the direction of pulling is made to coincide with the circumferential angle of the

DETAILED THREE-DIMENSIONAL BLADE DESIGN   169

tenth coordinate point of the mean streamline. In the meridional (r–z) plane the direction of pulling is defined by the tangent of the angle with the axial direction TNA. The direction of pulling should be approximately perpendicular to the mean streamline at the tenth coordinate point. For pure radial flow, the direction of pulling would be in direction of the z axis, the axis of rotation; therefore, TNA $= 0.0$. All blades of a radial impeller can be pulled together in a single action. The mean streamline direction does not need to be radial for the blades to be pulled together in the axial direction.

In the case of the design example considered here, the mean meridional streamline was chosen to be a straight line inclined from the radial direction. Coordinate points were defined by subdividing the distance between inlet and exit in 20 equal increments, 0.1375 in. (0.35 cm) in the r direction and 0.077 in. (0.2 cm) in the z direction. The inclination of the direction of pulling to the z axis was selected as TNA $= \tan 20°$.

The computer program also allows a choice of angular momentum distribution on the mean streamline (Fig. 12.5). A sinusoidal distribution is preferred which starts and ends gradually, in order to unload the blade leading and trailing edges. An exponent of the distribution function XN can modify the shape of the curve. An exponent smaller than 1 increases blade loading toward the front; an exponent greater than 1 increases loading toward the end of the blade.

$$\frac{C_t r}{C_{t2} r_2} = \left\{ 0.5 \left[1 - \cos\left(\frac{\pi I}{20}\right) \right] \right\}^{XN}$$

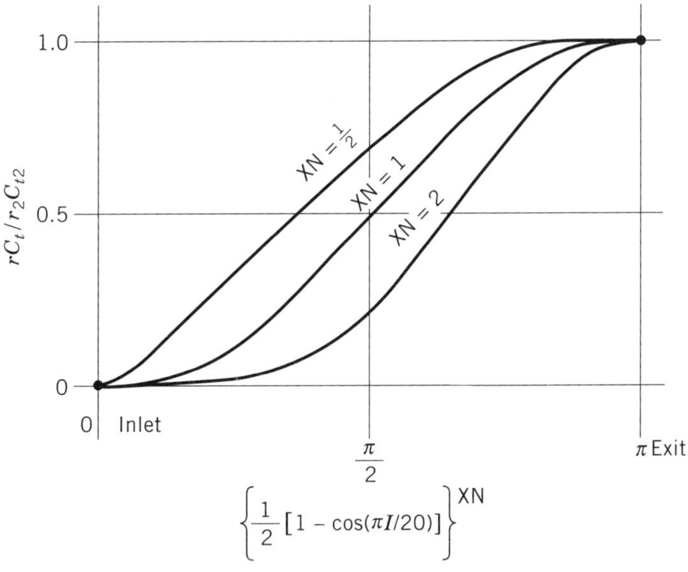

Figure 12.5 Angular momentum distribution function from pump inlet to exit.

170 DESIGN EXAMPLES

The printout of the computer program MIXED (Fig. 12.6), executed with an exponent XN = 1.0 and with the input above, shows (1) the three polar blade coordinates r, z, and θ for each streamline, the shroud or tip, the mean, and the hub, (2) the angular momentum $C_t r$ distribution along the three streamlines, and (3) the local blade angles β. The following observations can be made:

```
John Tuzson & Associates, Evanston,IL 8/25/99

Mixed flow impeller blade coordinates

   RT      ZT     THET       RM      ZM     THEM      RR      ZR    THER
 3.441   0.796   8.082    3.050   1.870   0.000    2.659   2.944 -13.798
 3.548   0.958  12.806    3.187   1.947   6.956    2.827   2.936  -1.449
 3.673   1.068  18.177    3.325   2.024  13.789    2.977   2.980   7.951
 3.799   1.176  23.675    3.463   2.101  20.416    3.126   3.026  16.282
 3.926   1.283  29.153    3.600   2.178  26.774    3.274   3.073  23.854
 4.053   1.388  34.503    3.737   2.255  32.820    3.422   3.122  30.802
 4.181   1.490  39.656    3.875   2.332  38.529    3.569   3.174  37.202
 4.310   1.591  44.566    4.012   2.409  43.890    3.715   3.227  43.104
 4.440   1.688  49.212    4.150   2.486  48.905    3.860   3.284  48.552
 4.571   1.784  53.588    4.287   2.563  53.588    4.004   3.342  53.588
 4.703   1.877  57.703    4.425   2.640  57.962    4.147   3.403  58.256
 4.835   1.968  61.580    4.562   2.717  62.061    4.290   3.466  62.604
 4.968   2.056  65.247    4.700   2.794  65.923    4.432   3.532  66.683
 5.103   2.143  68.742    4.838   2.871  69.594    4.572   3.599  70.550
 5.237   2.227  72.109    4.975   2.948  73.125    4.713   3.669  74.261
 5.373   2.309  75.394    5.113   3.025  76.567    4.852   3.741  77.878
 5.509   2.390  78.647    5.250   3.102  79.977    4.991   3.814  81.464
 5.646   2.468  81.917    5.387   3.179  83.412    5.129   3.890  85.084
 5.784   2.544  85.254    5.525   3.256  86.927    5.266   3.968  88.803
 5.923   2.618  88.707    5.662   3.333  90.580    5.402   4.048  92.688

         Angular Momentum              Blade Angles

  6.750    0.033   -5.855    56.439    67.842   74.526
  4.393    0.353   -3.537    64.148    68.327   72.302
  4.435    0.883   -2.098    65.477    68.519   71.085
  4.802    1.613   -0.890    66.181    68.471   70.213
  5.445    2.525    0.287    66.455    68.220   69.445
  6.321    3.599    1.505    66.415    67.797   68.681
  7.385    4.807    2.787    66.140    67.229   67.878
  8.597    6.122    4.130    65.690    66.546   67.027
  9.915    7.510    5.519    65.119    65.785   66.140
 11.297    8.938    6.930    64.485    64.993   65.251
 12.704   10.371    8.336    63.850    64.226   64.409
 14.095   11.772    9.704    63.286    63.554   63.679
 15.432   13.108   11.003    62.869    63.051   63.135
 16.680   14.344   12.200    62.673    62.792   62.851
 17.807   15.451   13.264    62.761    62.840   62.894
 18.785   16.400   14.166    63.171    63.237   63.305
 19.591   17.167   14.880    63.910    63.988   64.092
 20.211   17.734   15.381    64.950    65.067   65.228
 20.637   18.085   15.645    66.231    66.413   66.658

       53.58777618      20.00000000
```

Figure 12.6 Input and output files of the computer program MIXED, calculating blade coordinates for a mixed flow pump impeller design example having parallel straight-line blade surface generator lines.

1. An obvious but easily overlooked requirement is that the sense of rotation of the impeller be correct. The sense of the blade can be adjusted by counting the circumferential angle θ either clockwise or counterclockwise.
2. The circumferential wrap angles, about 80° on the shroud and about 110° on the hub, are reasonable. At the inlet the blade is inclined with respect to the meridional plane but is only slightly slanted at the exit since the circumferential angles on hub and shroud differ by only 4°.
3. The angular momentum distribution shows that there is a positive incidence at the shroud and a negative incidence at the hub. At the exit about 15% more energy is added to the fluid on the shroud and 15% less on the hub. Leading- and trailing-edge adjustments would be needed if equal-energy addition on all streamlines were desired. The streamline curvature at the inlet will alleviate the incidence problem somewhat. The meridional inlet velocity will be higher than the average at the shroud and lower at the hub, which will reduce the incidence in both locations. A detailed inlet flow calculation should be made at this point to determine the actual meridional velocities on hub and shroud. Local blade angle changes can be made by thinning the blade on the pressure or suction side. At a blade angle of 68°, a 20/1000 in., thickness change over a length of 1 in. (0.5 mm over 25 mm) corresponds to a 1° change in the blade angle. Actual blade angles are very difficult to measure, especially since the effective angle should be taken along the streamline, the exact direction of which is usually unknown. The front and exit ends of the blade can be adjusted on the hub by adding coordinate points. If the blade is pulled toward the hub, additional blade length on the hub does not affect the shroud.
4. It is important to remember that the blade angle β is measured from the meridional direction, not the radial direction.

The blade curve shown on a projection onto a plane perpendicular to the axial direction, the $r-\theta$ plane, as customary on engineering drawings, does not form an angle β with the radial direction, but an angle β_r. The following relationships exist between the meridional blade increment Δm, the radial increment Δr, and the circumferential increment $r\Delta\theta$:

$$\tan \beta = \frac{r\Delta\theta}{\Delta m}$$

$$\cos \alpha = \frac{\Delta r}{\Delta m}$$

$$\tan \beta_r = \frac{r\Delta\theta}{\Delta r} = \frac{\tan \beta}{\cos \alpha}$$

$$\Delta\theta_{\text{radians}} = \Delta\theta_{\text{degrees}}\left(\frac{\pi}{180}\right)$$

In these expressions the circumferential angle increment $\Delta\theta$ must be measured in radians, not degrees.

172 DESIGN EXAMPLES

5. Alternative blade designs can be executed with different angular momentum distributions XN, with different directions of pulling TNA, including an axial pull TNA = 0.0, and with different mean streamline coordinates. The closer the inclination of the blade surface generating lines to the axial direction, the less will be the difference in angular momentum on the streamlines. Ideally, the blades should deliver uniform angular momentum distribution at the impeller exit, but often a compromise will be necessary. The designer's judgment will decide whether the compromise is acceptable and satisfactory.

Plots of the blade coordinates calculated here, which are available in the file MIXED.GRF and can be transferred to a plotting program, show that the blade shapes are smooth and error free (Fig. 12.7). They can be made to

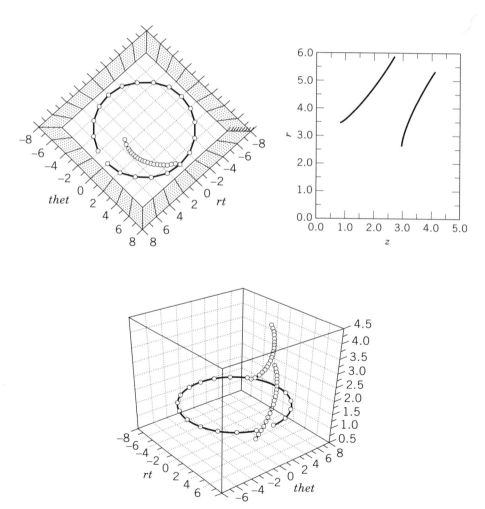

Figure 12.7 Plots showing impeller blade coordinates at the hub and shroud for the test case of Fig. 12.6.

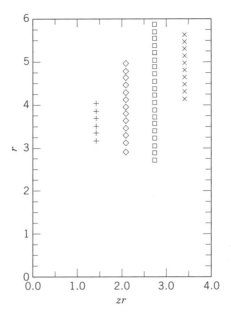

Figure 12.8 Sample printout of the computer program CUTS, calculating sections of blades in consecutive circumferential planes for the test case of Fig. 12.6.

overlap when rotated and viewed from the direction in which they would be pulled. For the purposes of preparing engineering drawings, the intersections of the blade surface with planes perpendicular to the z axis, r–θ planes, can be prepared with the help of the CUTS computer program. The input file CUTS.DAT, containing the blade coordinates on the hub and shroud in free form, can be prepared, for example, by editing the file MIXED.DTA from the MIXED computer program. The file needs to be renamed and the text deleted. Four cuts are calculated, equally spaced between the shroud inlet location and the hub exit. The coordinates r, z, and θ of the six cuts are listed in the output file CUTS.OUT (Fig. 12.8) and are suitable for transfer to a plotting program.

TWO-DIMENSIONAL BLADE DESIGN

At lower specific speeds, a two-dimensional blade shape becomes possible. The blade surface does not depend on the axial z coordinate and can be fully defined by the radial r and circumferential θ variables. Such a blade shape becomes possible if the flow velocities along the hub and the shroud are essentially the same. The blade coordinates can still be calculated with the MIXED computer program by making the blade surface generator lines parallel to the axial direction, and setting TNA = 0.0. The blade surface is formed with axial generator lines, and the blade can be pulled in the axial direction.

174 DESIGN EXAMPLES

After executing the MIXED program with the input as above and TNA = 0.0, the printout in Fig. 12.9 shows identical radial coordinates for the corresponding points of the three streamlines. A slight difference appears in the angular momentum distribution and in the blade angles because the meridional velocity changes from inlet to outlet and the hub and shroud streamlines are not parallel to the mean streamline in the meridional (r–z) plane view. The blade angles are measured along the streamlines, as before.

```
John Tuzson & Associates, Evanston,IL 8/25/99

Mixed flow impeller blade coordinates. Axial pull.

   RT       ZT      THET      RM       ZM      THEM      RR       ZR      THER
 3.050    0.796    0.000    3.050    1.870    0.000    3.050    2.944    0.000
 3.187    0.756    6.956    3.187    1.947    6.956    3.187    3.138    6.956
 3.325    0.873   13.789    3.325    2.024   13.789    3.325    3.175   13.789
 3.463    0.988   20.416    3.463    2.101   20.416    3.463    3.214   20.416
 3.600    1.101   26.774    3.600    2.178   26.774    3.600    3.255   26.774
 3.737    1.211   32.820    3.737    2.255   32.820    3.737    3.299   32.820
 3.875    1.319   38.529    3.875    2.332   38.529    3.875    3.345   38.529
 4.012    1.424   43.890    4.012    2.409   43.890    4.012    3.394   43.890
 4.150    1.526   48.905    4.150    2.486   48.905    4.150    3.446   48.905
 4.287    1.625   53.588    4.287    2.563   53.588    4.287    3.501   53.588
 4.425    1.721   57.962    4.425    2.640   57.962    4.425    3.559   57.962
 4.562    1.815   62.061    4.562    2.717   62.061    4.562    3.619   62.061
 4.700    1.906   65.923    4.700    2.794   65.923    4.700    3.682   65.923
 4.838    1.994   69.594    4.838    2.871   69.594    4.838    3.748   69.594
 4.975    2.080   73.125    4.975    2.948   73.125    4.975    3.816   73.125
 5.113    2.163   76.567    5.113    3.025   76.567    5.113    3.887   76.567
 5.250    2.244   79.977    5.250    3.102   79.977    5.250    3.960   79.977
 5.387    2.323   83.412    5.387    3.179   83.412    5.387    4.035   83.412
 5.525    2.399   86.927    5.525    3.256   86.927    5.525    4.113   86.927
 5.662    2.473   90.580    5.662    3.333   90.580    5.662    4.193   90.580

         Angular Momentum              Blade Angles

 -1.162    0.033    3.981    69.714   67.842   58.482
  1.914    0.353   -0.961    65.553   68.327   70.234
  2.437    0.883   -0.436    65.888   68.519   70.357
  3.115    1.613    0.323    65.997   68.471   70.238
  3.939    2.525    1.295    65.911   68.220   69.912
  4.899    3.599    2.452    65.657   67.797   69.407
  5.979    4.807    3.760    65.263   67.229   68.750
  7.156    6.122    5.185    64.758   66.546   67.969
  8.408    7.510    6.687    64.179   65.785   67.101
  9.704    8.938    8.228    63.571   64.993   66.189
 11.012   10.371    9.769    62.993   64.226   65.292
 12.299   11.772   11.272    62.511   63.554   64.477
 13.530   13.108   12.703    62.200   63.051   63.822
 14.671   14.344   14.029    62.133   62.792   63.402
 15.687   15.451   15.220    62.372   62.840   63.284
 16.548   16.400   16.254    62.954   63.237   63.510
 17.226   17.167   17.109    63.883   63.988   64.092
 17.697   17.734   17.771    65.128   65.067   65.006
 17.939   18.085   18.230    66.627   66.413   66.198

        53.58777618    0.00000000E-01
```

Figure 12.9 Printout of the computer program MIXED, calculating blade coordinates for a mixed flow pump impeller design example having axial blade surface generator lines.

```
John Tuzson & Associates, Evanston, IL. 4/16/95

BLADE - Impeller blade coordinate calculation

R = Radius, BETA = Blade angle from tangent
THETA = Angle at center of rotation from Y axis
RHO = Radius of curvature
XO,YO = Coordinates of center of curvature

     R      Beta     Theta      Rho      XO       YO

   3.250   22.000    0.000
   3.888   22.000  -24.931    3.849   -1.442   -0.319
   4.525   22.000  -46.177    4.537   -1.407   -1.005
   5.162   22.000  -64.679    5.224   -1.125   -1.633
   5.800   22.000  -81.060    5.912   -0.659   -2.138
```

Figure 12.10 Printout of the computer program BLADE, calculating radii and centers of curvature for defining by arcs of circles the blades of a radial flow pump impeller design example.

If the impeller flow passage and the streamlines are substantially radial, a very simple blade shape can be specified by executing the BLADE2 code, which defines the blade by segments of arcs of circles. The input file BLADE.DAT requires the inlet and outlet angles and five radii at which the arcs of circles join. The output, shown in Fig. 12.10, lists the radii, the blade angles, which are incremented from inlet to exit in proportion to the radial increments, the circumferential angle, the radius of curvature, and the coordinates of the centers of curvature of the arcs of circles. In this case the blade angles calculated are indeed the angles in the plane $(r-\theta)$, perpendicular to the axial direction.

The computer program assumes an axial, constant-radius blade leading edge. In reality the leading edge in practical impellers can be chosen inclined. The radius at the shroud inlet is larger than the radius at the hub inlet. Therefore, the blade surface should be calculated starting at the hub inlet radius, and the blade should be trimmed at the shroud. Since the blade angle at the hub inlet does not affect the shroud inlet, it can be chosen at will. However, when the radius of the blade surface reaches the radius of the shroud inlet, the blade angle must take on the value that will give correct incidence at the shroud inlet. As before, the leading edge of the blade needs to be adjusted. Even in purely radial impellers, the flow enters the impeller eye in the axial direction and turns toward the radial direction in the impeller. Therefore, the meridional velocities at the hub and shroud will not be the same. They must be calculated separately, and the blade angles at the inlet must be adjusted to give correct incidence. In low-specific-speed pumps, relatively low velocities exist at the impeller inlet, and incidence losses remain relatively low, which can be ascertained from the performance calculation program printout. Therefore, in this case, less ideal conditions can be tolerated at the impeller inlet.

RETURN PASSAGE DESIGN

Assuming that a return passage follows the mixed flow impeller (Fig.12.11) which was calculated previously, the computer program MIXED can again be

176 DESIGN EXAMPLES

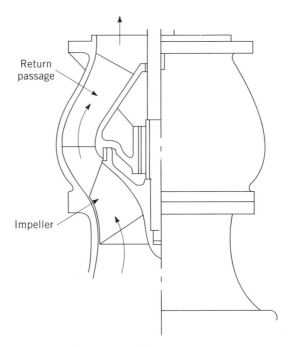

Figure 12.11 Return passage.

used to calculate the diffuser vane coordinates. The sense of direction of the calculation needs to be reversed. The return passage exit, at the entrance of the next stage, becomes the first station to be calculated with the positive direction of the axial, z coordinate pointing backwards with respect to the direction of the flow. The return passage inlet, near the impeller exit, becomes the last station to be calculated. The input variables must be assigned accordingly.

The coordinates of the mean streamline, r and z, which need to be input, will follow an S-shaped curve starting axially at the average radius of the impeller inlet, and ending again axially at the return passage radius D3/2 which has been determined in the performance calculation with the code LOSS3. The z-coordinate points correspond to 20 equal increments. The radial coordinate remains constant for two or three stations at the inlet and exit, and follows a constant slope, for example, 45°, in between. The slope changes need to be rounded off. The program MIXED smoothes the input data with a simple averaging over three stations.

As to the other input variables: the flow rate $Q = 2000$ gpm ($0.126\,\mathrm{m^3/s}$), rotational speed $N = 0.0$ rpm, the vane thickness $T = 0.25$ in. (6.35 mm), number of vanes XNB = 7, inlet meridional velocity WM1 = 19.34 ft/sec (5.98 m/s), and exit meridional velocity WM2 = 13.35 ft/sec (4.07 m/s) are available from the performance calculation. The absolute tangential exit velocity CT21 must be calculated from the impeller exit tangential velocity CT2, given in the printout of the performance program, by accounting for the increase in the radius from impeller exit to return passage inlet:

$$CT21 = CT2(D2/D3) = 40.42 \text{ ft/sec}(11.6 \text{ in.}/12.75 \text{ in.}) = 36.75 \text{ ft/sec}$$
$$CT21 = 12.32 \text{ m/s}(0.295 \text{ m}/0.328 \text{ m}) = 11.20 \text{ m/s}$$

The inclination of the direction, in which the vane would be pulled from the casting mold, GAMAREF = 45°, was chosen approximately perpendicular to the mean streamline. The exponent of the tangential velocity distribution XN = 3 will load the vanes more at the end, where the flow enters the return passage. No slip would be expected in stationary passages: SIGMA21 = 1.0.

The printout of the MIXED code and the plot of the r and z coordinates, with the foregoing input for the return passage in Figs. 12.12 and 12.13, may show that the first, trial meridional streamline input could have used some modifications at both vane ends. A generous fillet at the hub and large radius of curvature at the shroud would streamline the contours. Slight modifications would not be expected to affect the flow velocities, especially since boundary layers and blockage have been neglected. The vane wrap angle varying from 50 to 75° is reasonable but will change if the vane ends are trimmed. The tangential velocity changes gradually, and little difference exists from one streamline to the next. The local vane angles could use some smoothing, but all reach 70° at the end, where the flow from the impeller enters, in agreement with the value that was calculated with the performance program. The three-dimensional view shows that the hub and shroud streamlines indeed overlap if viewed from the direction in which they would be pulled.

VOLUTE DESIGN EXAMPLE

For the sake of illustrating the design procedure, the principal dimensions of a volute will be chosen for the impeller delivering 2000 gpm (0.126 m³/s) at 100 ft (30.5 m) head and rotating at 1770 rpm, designed previously. The impeller has an 11.6-in. (0.295-m) diameter D2 and an exit width of B2 = 1.3 in. (3.3 cm). The diameter of the base circle tangent to the volute tongue will be made about 10% larger than the impeller, about 12.75 in. (0.324 m). Pressure fluctuations at the tongue when the individual blades pass by will decrease with increasing base circle diameter.

The volute throat area is determined by matching the velocity coming from the impeller exit with the throat velocity calculated from the flow rate. At design conditions, $Q = 2000$ gpm (0.126 m³/s), the absolute tangential velocity leaving the impeller has been calculated: CT2 = 40.42 ft/sec (12.32 m/s). To calculate the volute throat cross-sectional area, let us assume for a first estimate that the velocity at the throat has slowed down because of the increased radius to

Return passage test case.

John Tuzson & Associates, Evanston, IL 8/25/99

Mixed flow impeller blade coordinates

RT	ZT	THET	RM	ZM	THEM	RR	ZR	THER
3.627	-1.327	0.847	2.300	0.000	0.000	0.973	1.327	-3.164
3.510	-0.820	0.765	2.350	0.340	0.000	1.190	1.500	-2.257
3.504	-0.397	0.711	2.428	0.680	0.000	1.351	1.757	-1.846
3.537	0.010	0.663	2.527	1.020	0.003	1.517	2.030	-1.537
3.592	0.420	0.618	2.652	1.360	0.017	1.712	2.300	-1.246
3.676	0.837	0.594	2.813	1.700	0.066	1.950	2.563	-0.929
3.812	1.264	0.636	3.036	2.040	0.207	2.260	2.816	-0.517
4.032	1.680	0.851	3.331	2.380	0.543	2.631	3.080	0.072
4.286	2.072	1.358	3.638	2.720	1.188	2.990	3.368	0.944
4.564	2.458	2.313	3.962	3.060	2.313	3.360	3.662	2.313
4.855	2.835	3.881	4.290	3.400	4.088	3.726	3.965	4.357
5.151	3.204	6.218	4.615	3.740	6.672	4.079	4.276	7.246
5.443	3.566	9.442	4.929	4.080	10.191	4.415	4.594	11.115
5.741	3.923	13.728	5.245	4.420	14.826	4.748	4.917	16.159
6.052	4.279	19.203	5.571	4.760	20.711	5.090	5.241	22.521
6.350	4.630	25.615	5.880	5.100	27.602	5.410	5.570	29.978
6.592	4.968	32.300	6.120	5.440	34.884	5.648	5.912	37.997
6.773	5.280	38.845	6.273	5.780	42.309	5.773	6.280	46.613
6.913	5.556	45.084	6.349	6.120	49.996	5.784	6.684	56.563
7.021	5.808	50.909	6.368	6.460	58.095	5.716	7.112	69.426

Angular Momentum			Blade Angles		
0.057	0.000	-0.086	-0.553	0.000	3.881
0.042	0.000	-0.052	-0.442	0.002	1.831
0.039	-0.001	-0.047	-0.420	0.018	1.468
0.036	-0.006	-0.053	-0.389	0.101	1.494
0.019	-0.024	-0.068	-0.209	0.371	1.745
-0.032	-0.070	-0.100	0.359	1.051	2.324
-0.166	-0.172	-0.165	1.839	2.483	3.394
-0.435	-0.385	-0.324	4.640	5.111	5.654
-0.893	-0.768	-0.627	9.089	9.407	9.642
-1.629	-1.401	-1.155	15.579	15.698	15.636
-2.719	-2.361	-1.987	23.952	23.882	23.592
-4.216	-3.711	-3.189	33.388	33.200	32.815
-6.126	-5.450	-4.760	42.682	42.432	42.021
-8.417	-7.566	-6.703	50.758	50.521	50.170
-11.014	-10.052	-9.089	57.057	56.989	56.898
-13.725	-12.804	-11.925	61.615	61.856	62.216
-16.341	-15.456	-14.649	65.007	65.359	65.872
-18.788	-17.624	-16.639	67.676	67.763	68.073
-20.163	-19.026	-19.712	68.993	69.275	71.337

2.31322575   45.00000000

Figure 12.12 Printout of the computer program MIXED, calculating the vane coordinates of the return passage of a mixed flow pump impeller design example.

VOLUTE DESIGN EXAMPLE 179

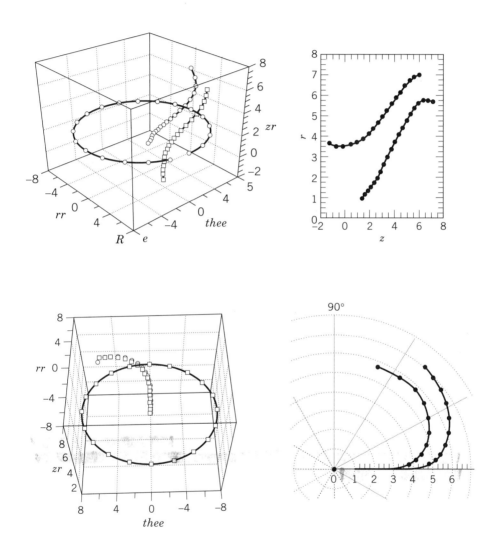

Figure 12.13 Plots showing the hub and shroud coordinates of the return passage vanes of Fig. 12.12.

CQ3 = 36.75 ft/sec (11.20 m/s). The throat area A, assumed to be circular with diameter d, would then be calculated from

$$A = \frac{\pi d^2}{4} = \frac{Q}{CQ3} = \frac{2000 \times 231/60}{36.75 \times 12} = 17.46 \text{ in}^2$$

$$A = \frac{0.126 \text{ m}^3/\text{s}}{11.20 \text{ m/s}} = 0.01125 \text{ m}^2 = 112.5 \text{ cm}^2$$

Assuming a circular volute flow passage cross section, the throat area diameter will be $d = 4.7$ in. (12.0 cm). However, the center of the throat area will move to a radius of (base circle diameter $+ d$)/2, and the velocity approaching the throat will slow down in inverse proportion to the distance from the impeller center. Therefore, the throat velocity will diminish to

$$CQ3 = CT2[D2/(D3 + d)] = 40.42[11.6/(12.75 + 4.7) = 26.87 \text{ ft/sec}$$
$$CQ3 = 12.32 \text{ m/s}[0.295 \text{ m}/(0.324 \text{ m} + 0.120 \text{ m}) = 8.20 \text{ m/s}$$

Therefore, the corrected area will become

$$A = \frac{\pi d^2}{4} = \frac{Q}{CQ3} = \frac{2000 \times 231/60}{26.87 \times 12} = 23.88 \text{ in}^2$$

$$A = \frac{0.126 \text{ m}^3/\text{s}}{8.2 \text{ m/s}} = 0.01536 \text{ m}^2 = 154 \text{ cm}^2$$

Depending on the actual shape of the throat cross section, which could approach the rectangular or trapesoidal, and the location of the average streamline, slightly different throat areas would be calculated. The calculated area corresponds to an effective flow cross section. In practice, the geometrical area should be 15 to 25% greater because of boundary layer build up in the volute and because wall friction and sudden expansion losses tend to slow the velocity. A larger volute throat will tend to shift the best efficiency point to higher flow rates and can reduce diffusion losses at the pump exit, which, however, will depend on the effectiveness of any diffuser, which may follow the volute.

The volute flow cross sections at other locations around the circumference of the impeller vary in proportion to the circumferential angle θ measured from the tongue:

$$\text{Cross-sectional area at } \theta = (\text{throat area})\left(\frac{\theta}{360}\right)$$

The width of the volute in the axial direction immediately at the impeller exit should ideally be slightly larger than the impeller exit width, with close clearances between the impeller sides and the volute walls. However, high efficiencies have been reached with much wider widths. An obvious but easily overlooked issue in volute design is the sense of rotation of the impeller.

Mechanical design considerations strongly influence volute design. The volute passages, being under high hydraulic pressure, exert strong forces on the casing structure, and some of the passage walls remain weakly supported. Industrywide standards often prescribe the location of the inlet and exit flanges, mounting bolts, and external loads, which the bolts have to support. Manufacturing, assembly, and installation consideration also come into play.

CONCLUDING REMARKS

The simple design tools offered here—the computer programs LOSS3, INLET, IMPEL, MIXED, CUTS, BLADE2—can lead to a first trial design, which can then be refined at will. The data may serve as input to more detailed and refined calculation procedures and computer programs. The degree of refinement, and the time spent on them, will be governed by the value of the pump. Hydraulic design is only a modest part of the overall design effort. It must be supplemented by rotor dynamic and stress analysis calculations, and finally, by a manufacturing cost analysis. These other requirements may impose constraints on the hydraulic design and make a new hydraulic design iteration necessary.

13
ROTOR DYNAMICS OF PUMPS

FUNDAMENTALS OF SHAFT VIBRATIONS

A general overview of pump rotor dynamics can be found in the books of Vance (1988), Childs (1998), and the publication of Corbo and Malanoski (1998), for example. Only a brief overview is given here.

Any mass held in position elastically is capable of oscillations if its position is disturbed. Because of force balance, the inertia force, the mass m times acceleration, must be balanced by the spring force, the spring rate k times the displacement. The spring constant k has the dimension of force per unit length. On dimensional grounds the frequency of oscillation f must be given by the following proportionality, the constant of proportionality depending on the detailed configuration of the system:

$$f : \left(\frac{k}{m}\right)^{1/2}$$

When an outside excitation has the same frequency as the natural frequency of the system, resonance is encountered. The amplitude of oscillations increases progressively. The maximum amplitude can be limited by friction or some other energy-dissipating force that extracts energy from the oscillating system. Operation at, or passage through, the critical speed becomes possible if sufficient damping maintains deflections sufficiently low (Barrett et al. 1978). Otherwise, the amplitude of oscillations will increase until the system destroys itself.

In the simplest case of a rotating system, the lateral deflection of the shaft will result in a restoring spring force, and the mass of the rotor, usually much greater than the mass of the shaft, will stand for the mass in the expression above (Childs 1993, Vance 1988). The rotational frequency, $\omega/2\pi$, will excite the oscillation. The system reaches its critical speed when the rotational frequency equals the natural frequency of the system. The magnitude of the exciting force will be in proportion to the unbalance of the rotor. Ideally, if rotors could be perfectly balanced, no excitation could take place. Unfortunately, some unbalance always remains. Standards by the Hydraulic Institute, ISO, ASME, and API prescribe the permissible unbalance of various rotating assemblies.

Static balancing can be accomplished by placing the rotating assembly on knife edges or by mounting on low-friction bearings, and by letting the assembly rotate to its equilibrium position under the effect of gravity. The direction of unbalance, which is the eccentric position of the center of gravity of the rotor with respect to the axis of rotation, then points downward. Adding small weights, by trial and error, in the opposite, upper position, a balanced, neutral equilibrium state can be reached in which the rotor remains in any angular position without moving.

When the rotating mass is long, an unbalance moment can still remain from eccentric mass distributions on its ends, which tend to produce a longitudinal bending torque on the shaft. Such unbalance moments can also be produced by large-diameter rotors, which are not mounted perpendicular to the rotational axis or by hydrodynamic forces on wobbling impellers. Balancing machines and procedures exist in industry to detect and correct not only static unbalance, but also unbalance torques resulting from eccentric mass distributions. Shaft bow or distortion can also excite oscillations, and these act in a manner similar to unbalance. Depending on the relative directions of bowing and unbalance, their combined effect can give rise to responses that can be different from those of unbalance only.

For the purposes of an analysis, the instantaneous transverse deflections of the shaft should be separated into two components, corresponding to the two perpendicular directions of the force components and displacement variables. Ideally, vibrations in the two directions are independent. However, hydrodynamic forces in bearings and in close clearances, such as seals, introduce a coupling between the two directions which must therefore be considered jointly. Indeed, the deflection in a sleeve bearing, which produces an eccentric position of the shaft with respect to the bearing bore, results in a bearing force that is not directed opposite to the deflection as an ordinary spring would, but also produces a force component in the transverse direction. Bearing supports, on the other hand, can deflect elastically, producing forces directly opposed to, and proportional to, the deflection. Such a support flexibility is equivalent to greater flexibility of the shaft.

A further complication is introduced by the fact that complex rotating systems, having several rotors or long shafts, for example, have more than one critical speed. The system has several modes or configurations of oscillation, each having a different natural frequency. Such natural frequencies can be detected on existing assemblies with vibration-measuring instruments but are difficult to predict exactly during the design of the pump assembly. The numerical rotor dynamic analysis of such complex systems requires the use of modern computer-aided calculations.

In pump design and operation, shaft vibrations affect two issues. Excessive shaft deflections may cause metal-to-metal contact in bearings, wear rings, or close clearances that could damage the pump. Excessive shaft oscillations may generate vibrations that may overload the bearings—rolling element bearings also—and may even cause structural damage. Consequently, it is important to estimate the shaft deflections and vibration amplitudes when designing pumps. Shaft oscillation at or near the critical speed can be limited by sufficient damping at the bearings or their support (Barrett et al. 1978).

Other important mechanisms exciting oscillations below the first critical speed are intermittent, once per revolution, rubbing, or loss of contact at bearings or seals (Childs 1993). The oscillations appear at an exact fraction of the running speed, usually one-half but at times one-third or one-fourth of the speed. Ball bearings, for example, remain stiff as long a contact is maintained between the balls and the races. However, unless the bearings are preloaded, some clearance always exists, and contact can be lost temporarily and periodically during each revolution.

An important distinction must be made between shaft deflections that are synchronous, always pointing in the same direction with respect to the rotating shaft, and deflections that rotate with respect to the shaft. Unbalance and shaft bowing excite synchronous deflections. Oscillations excited by other instabilities, such as rubbing, are not synchronous. The corresponding deflections rotate with respect to the shaft, flex the shaft periodically, and create alternating bending stresses in the shaft. Internal friction can be produced between the bending shaft and an impeller pressed onto the shaft, which can also affect shaft dynamics (Childs 1993).

Several mechanisms can cause instability and excite oscillations at frequencies below the rotational frequency of the shaft (shaft whirl) if the assembly is operated above its first critical speed (Ehrich and Childs 1984). Such might be the case for high-speed boiler feed pumps or rocket fuel pumps but are less likely to be encountered in general industrial applications.

SHAFT AND IMPELLER SUPPORTED ON TWO BEARINGS

An impeller on a shaft, supported by two bearings, constitutes the simplest configuration, which is encountered frequently. Two cases are illustrated here (Fig. 13.1): an overhung impeller corresponding to an end suction pump, and an impeller and shaft assembly supported by bearings on either side, corresponding to a double suction pump. Ideally, the bearings can be assumed rigid, which may be the case for preloaded ball bearings, for example. If, in addition, the effect of the front wear ring forces and the damping effect of the water surrounding the impeller is neglected, the system can be modeled by the concentrated mass of the

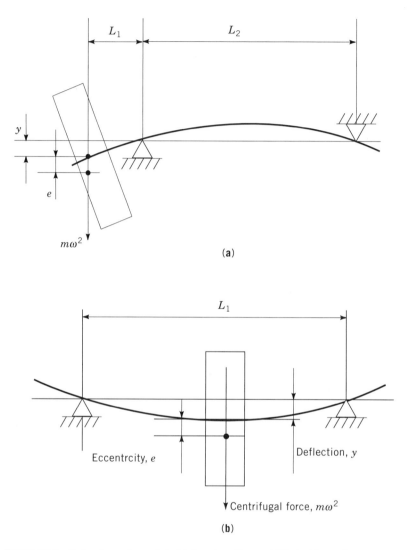

Figure 13.1 Rotor dynamic model of (a) end suction and (b) double suction pumps.

impeller, including the water contained in it, and the bending spring rate of the shaft. A first, approximate estimate of the natural frequency of the impeller and shaft assembly can be calculated. Usually, the natural frequency is well above the frequency of the shaft rotation. If the shaft were of uniform cross section, the shaft spring rate for overhung impellers would be given by

$$k = \frac{3EI}{L_1^2(L_2 + L_1)}$$

$$I = \frac{\pi}{64} d^4$$

In this expression, E is the elastic modulus of the shaft material, I the moment of inertia of the shaft, L_1 the overhang, L_2 the distance between bearings, and d the shaft diameter.

For impellers supported by bearings on either side, for example for double suction pumps,

$$k = \frac{48EI}{L_1^3}$$

If the water-filled impeller mass is m, the natural frequency f_n will be given by

$$f_n = \frac{1}{2\pi}\left(\frac{k}{m}\right)^{1/2}$$

Of perhaps greater importance than the critical speed will be the deflection at the front wear ring at the nominal shaft speed. It is assumed that an unbalance is present, which acts as if the center of gravity of the impeller were at a small distance e from the axis of rotation. The mass of the impeller, assumed to be concentrated at its center of gravity, rotates eccentrically and results in a radial centrifugal force resisted by the spring force of the shaft. Well below the critical speed, the direction of the centrifugal force is in the direction of the eccentricity e and the deflection y, and rotates synchronously, always pointing in the same direction with respect to the shaft. The deflection at the impeller can then be calculated from the equality of the radial centrifugal force and the spring force of the shaft. The centrifugal force is $m\omega^2(e+y)$, and the resisting spring force is ky. The following expression gives the deflection:

$$y = \frac{m\omega^2}{(k - m\omega^2)}e = \frac{\omega^2}{(\omega_c^2 - \omega^2)}e$$

$$\omega_c^2 = \frac{k}{m}$$

It can be observed that if the rotational speed ω becomes equal to the critical speed ω_c, the deflection becomes infinite, assuming that no damping is present (Fig. 13.2). If the shaft stiffness k is much greater than $m\omega^2$—the rotational speed remains far below the critical speed—the expression simplifies to

$$y = \frac{m}{k}\omega^2 e = \frac{\omega^2}{\omega_c^2}e$$

The deflection is proportional to the unbalance eccentricity and the ratio of the rotational speed and the critical speed squared.

The deflection at the front wear ring may be slightly larger or smaller than the deflection at the impeller center of gravity, where y is measured, depending on the

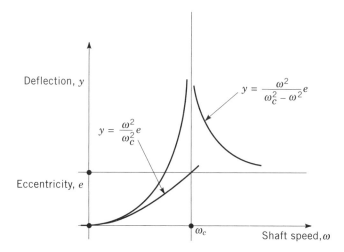

Figure 13.2 Shaft deflection due to unbalance as a function of speed for simple rotor dynamic models.

spacing of the bearings and location of the impeller. Evidently, the clearance of the wear ring must be larger than the shaft deflection. Large deflections and large clearances are undesirable because they result in large leakage.

The numerical values given by the expressions above approximate only rarely the real values and give at best a very rough estimate. Sleeve bearings deflect under load and add to the shaft flexibility. Hydrodynamic bearings and seals introduce other force components which change the ideal values of the critical speeds and deflections. An impeller that is not mounted perpendicular to the shaft, or fluid forces resulting from a wobbling impeller, introduce twisting moments on the rotating assembly that depend on the rotational speed and influence the critical speed of the pump.

REFERENCES

Barrett, L. E., Gunter, E. J., Allaire, P. E. (1978): Optimum Bearing and Support Damping for Unbalance Response and Stability of Turbomachinery, *ASME Journal of Engineering for Power*, January, pp. 89–94.

Childs, D. (1993): *Turbomachinery Rotordynamics*, Wiley, New York.

Corbo, M. A., Malanoski, S. B. (1998): Pump Rotordynamics Made Simple, *Proceedings of the 15th International Pump Users Symposium*, Texas A&M University, College Station, Texas, pp. 167–204.

Ehrich, F., Childs, D. (1984): Identification and Avoidance of Instabilities in High Performance Turbomachinery, *Mechanical Engineering*, May, pp. 66–79.

Vance, J. (1988): *Rotordynamics of Turbomachinery*, Wiley, New York.

14

INLET AND EXIT RECIRCULATION

RECIRCULATION PHENOMENA

Pump users are particularly concerned about the minimum flow rate that can be tolerated in a pump without risking damage. Reduced flow can result in pressure fluctuations, cavitation, and overheating (Cooper 1988; Gopalakrishnan 1988; *La Houille Blanche* 1980, 1982, 1985; Vlaming 1989). Exit recirculation can also influence the head–flow curve of the pump near shutoff and affect stability. Therefore, these phenomena deserve great attention.

At off-design conditions, especially at flow rates less than those for which the pump was designed, the flow often does not follow the impeller passage wall surfaces, and different flow patterns appear. Three flow phenomena can be distinguished more frequently: inlet recirculation, separation in the impeller, and exit recirculation. As the name indicates, in the case of inlet recirculation the flow, which already entered the impeller, turns around and flows back into the inlet pipe. In the case of exit recirculation, the flow, which has left the impeller, reverses and flows back into the impeller.

These phenomena have long been observed and studied (Guelich et al. 1993; Jansen 1967; Pfleiderer 1961; Sen 1979; Spannhake 1934; Von Karman Institute 1978). An understanding of the underlying causes was made difficult by the fact that these phenomena often appear combined, simultaneously, and interact. Complex interactions and instabilities may appear (Kaupert et al. 1996). However, as will be seen, their direct causes reside in the local geometry of the impeller. They can appear separately. Preferably they should be modeled independently from each other.

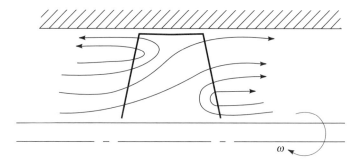

Figure 14.1 Inlet and exit recirculation flow patterns in an axial flow impeller at a reduced flow rate.

Inlet and exit recirculation become clearly evident in axial flow impellers or blade rows, as shown in Fig. 14.1. At reduced flow rates the main flow enters the blade row near the hub and follows an S-shaped path from the inlet near the hub to the exit near the periphery, where it leaves. Part of the entering flow splits off, turns around, and flows back into the inlet pipe near the periphery. Part of the flow, having left the blade row, turns around in the exit pipe and reenters the impeller near the hub. In centrifugal pumps, recirculation is most likely to appear, when the inlet diameter is large compared with the impeller diameter. Such proportions exist, in particular, in high-specific-speed mixed flow pumps, an extreme case being, precisely, axial flow impellers.

From an application engineering point of view, it has been desirable for pump users to derive statistically valid practical guidelines for the appearance of recirculation in typical commercially available pumps. Recirculation can result in severe flow oscillations and cavitation, which might erode the impeller, shorten bearing life, and reduce reliability. Such guidelines have been established by surveying typical pump populations (Fraser 1981). However, these empirical correlations do not establish a cause-and-effect relationship between specific geometrical features of the pump and recirculation. Consequently, they are not suited to provide guidelines for design and do not help to calculate the effect of recirculation on pump performance.

Although the actual phenomena are complex, simplified models are presented here that will allow their effect on pump performance to be calculated. A connection will be made between some features of the pump geometry and the appearance of these flow patterns.

INLET RECIRCULATION

To develop and illustrate a calculation model for inlet recirculation, let us assume for the time being that the leading edge is radial, and that at design flow rates the inlet blades are perfectly aligned with the direction of the inlet flow (Tuzson

1983). When the flow rate is reduced, separation appears on the suction side of the blades. The stagnant fluid in the separated region rotates with the impeller and is subject to the centrifugal acceleration. The static pressure in the separated region increases radially in proportion to the square of the radius:

$$\frac{p}{\rho} = \frac{\omega^2 r^2}{2}$$

This pressure also prevails on the separation streamline. Since the pressure remains constant at a given radius along the separated region, the relative velocity along the separated surface must also remain constant, as can be seen from the rothalpy equation if the foregoing value of the pressure is substituted:

$$\frac{P_1}{\rho g} + \frac{C_1^2}{2g} = \frac{p}{\rho g} + \frac{W^2}{2g} - \frac{\omega^2 r^2}{2g} = \frac{W^2}{2g}$$

The relative velocity is the vectorial sum of the tangential and meridional velocity components:

$$W^2 = C_m^2 + \omega^2 r^2$$

Since the quantity $\omega^2 r^2$ increases with the radius and the relative velocity W remains constant, the meridional velocity C_m must decrease with the radius. The separated region becomes progressively wider with increasing radius, and gradually blocks the inlet. Eventually, the point can be reached when the meridional velocity becomes zero at the greatest radius, at the shroud. This occurrence signals the onset of inlet recirculation.

For most practical pumps this flow rate is slightly below the design flow rate. However, slight recirculation near the shroud has little effect on performance. As will be seen, the parasitic power drain on the pump increases with a higher power of the decreasing flow rate and becomes significant only at generally lower flow rates. Because of the gradual appearance of inlet recirculation, it is difficult to designate a very specific flow rate below which dangerous pump deterioration can occur.

Once inlet recirculation has set in, the flow pattern in the pump inlet changes completely. The flow entering the impeller near the hub can still be assumed uniform and of equal energy. Such an assumption is supported by detailed measurements in the inlet of several mixed flow pumps (Hureau et al. 1993; Tanaka 1980; Tuzson 1983). It is also assumed that the meridional flow component in this portion of the inlet flow retains the magnitude, which will give correct incidence. Therefore, it is not the inlet velocity that is reduced when the flow rate becomes smaller but the effective cross-sectional area of the inlet. The remaining geometrical cross section of the inlet is taken up by the recirculating flow. Figure 14.2 illustrates such a flow pattern for an impeller with a slanted leading edge.

192 INLET AND EXIT RECIRCULATION

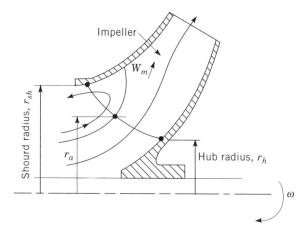

Figure 14.2 Inlet recirculation in a centrifugal pump at reduced flow rate. (From Tuzson 1983.)

If in the central portion of the inlet the same inlet velocity C_1 is assumed along a straight, slanted leading edge, the outer radius r_a of the flow, which passes through the impeller, is given by the relationship

$$r_a^2 = r_{sh}^2 - \frac{A}{\pi}\left(1 - \frac{Q}{Q_o}\right)$$

$$A = \pi\left(r_{sh}^2 - r_h^2\right)$$

where r_{sh} is the shroud radius, r_h the hub radius, and Q_0 the flow rate at design conditions. The remaining portion of the inlet, where the flow recirculates, can be regarded as a separate pump with an exit near the shroud and an inlet at the smaller radius r_a. The pressure Δp and flow rate q of such an imaginary pump would be given by expressions of the form

$$\Delta p \div \frac{\omega^2}{2}\left(r_{sh}^2 - r_a^2\right)$$

$$q \div \omega\left(r_{sh}^2 - r_a^2\right)^{3/2}$$

These expressions must have the form shown because of dimensional reasons, as can easily be verified. The axisymmetic geometry demands that the radii appear squared.

The expression for the power absorbed by recirculation will be proportional to the product of the pressure and the flow:

$$P_r = \Delta p\, q = C\rho \frac{\omega^3}{2}\left(r_{sh}^2 - r_a^2\right)^{5/2} = C\rho \frac{\omega^3}{2}\left(\frac{A}{\pi}\right)^{5/2}\left(1 - \frac{Q}{Q_0}\right)^{5/2}$$

This relationship is obtained by substituting the expression for r_a given above. The incremental head due to recirculation, ΔH_r, which would have to be added to the theoretical head of the pump when the efficiency is calculated is obtained by dividing the power of recirculation by the mass flow rate:

$$\Delta H_r = \frac{P_r}{\rho g Q} = C \frac{\omega^3}{2}\frac{1}{gQ}\left(\frac{A}{\pi}\right)^{5/2}\left(1 - \frac{Q}{Q_0}\right)^{5/2}$$

Unfortunately, the coefficient of proportionality C depends not only on the geometry of the pump, but also on the inlet piping configuration (Guesnon and Lapray 1985). For example, the flow returning into the inlet pipe can impart rotation to at least a portion of the flow reentering the pump. It will make a difference whether the pump is connected directly to a large tank or is preceded by a long pipe. The most practical method to determine the numerical value of the coefficient is to match the performance of an existing similar pump with the performance calculation computer program. The computer program LOSS3 contains a default value of 0.0005 for the coefficient, which implies certain conversion factors and should be verified from case to case.

This model of inlet recirculation leaves the flow through the impeller well behaved. The flow enters without incidence. The average relative velocity at the inlet W_1 remains constant. However, flow separation in the impeller can still appear at some point with reduced flow rate. If the streamline through the impeller from inlet to exit is relatively short, inlet recirculation can affect the exit flow conditions. Especially in mixed flow pumps, if exit recirculation also appears, the two phenomena can merge and result in very complex flow patterns.

Advanced impeller flow computer programs can calculate recirculating flow, but sometimes require that special inlet conditions be imposed (Graf 1993). Normally, uniform parallel flow is assumed upstream of the pump as the inlet boundary condition. Provision must be present for modeling leading-edge separation and boundary layer development. Some powerful programs incorporating advanced turbulence models can successfully calculate recirculating flow (Gopalakrishnan et al. 1995).

It will be noted that the proposed expression assumes that recirculation begins immediately below the design flow rate. In reality, the onset of recirculation depends strongly on the detailed design of the inlet, inlet blade angle distribution, location of the leading edge of the blades, and blade loading near the inlet. For example, if the direction of the flow meets the leading edge at a slant, a swept-wing effect may appear. The separated region will consist of a leading-edge vortex trailing along the leading edge, which will be swept along by the velocity component parallel to the leading edge. The presence of such a vortex can change entirely the conditions in the separated region, where no stagnant fluid will remain. The leading-edge vortex may convey the low momentum fluid of the separated region downstream, through the pump, and may retard the appearance of recirculation. The simple expression presented above cannot be expected to account for such details of the flow. Much more detailed geometrical data input would be needed to estimate the effect of leading-edge vortices.

Entirely different conditions also exist when substantial cavitation takes place. If the separated region behind the leading edge of the blades contains vapor bubbles, its bulk density will be low and the pressure will not increase in the radial direction along the leading edge. On the boundary—the separated streamline or stream surface—of vapor-filled separated regions, it will be the absolute velocity that will remain constant. In the case of intermittent cavitation, the flow pattern may switch back and forth between the normal flow configuration and a recirculating flow pattern. Violent flow oscillations would be expected and would be reflected by surges in power demand. These are the conditions that must be avoided in pump operation. These dangers explain the particular concern of pump operators with recirculation.

EXIT RECIRCULATION

Exit recirculation, when the flow returns into the impeller, appears at reduced flow rates in mixed flow pumps, which often have a slanting exit. Exit recirculation has been recognized as one of the phenomena limiting the operation of pumps at reduced flow rates (Gopalakrishnan 1988; Guelich et al. 1993). Determining its onset is of great interest to pump users.

The underlying cause of exit recirculation has given rise to much speculation. Measurements and visualization of the flow at the impeller exit show confused, unsteady fluctuations and strong interactions with the diffuser vanes or the volute, which are difficult to interpret. On the other hand, it is well known that ideal, axisymmetric, rotating shear flows—vortices, for example—can have stagnant or recirculating cores near the axis of rotation (Batchelor 1967; Keller and Egli 1985; Strscheletzky 1961; Tuzson 1993). The stability theory of rotating shear flows predicts that separate fluid regions of different energy can persist, with little mixing, if the low-energy fluid remains near the axis of rotation. One can therefore assume that a somewhat similar flow pattern can also exist at the pump exit when less energy is imparted to the fluid on the streamlines near the hub of the impeller. Ideal recirculating flow conditions are complicated in pumps by the unsteadiness of the flow and by the presence of diffuser vanes or the volute. Evidently, flow separation in the impeller can result in uneven angular momentum distribution at the impeller exit and can also precipitate exit recirculation (Jaberg and Hergt 1989). Still, the theoretical concept of nonuniform angular momentum distribution can serve as a basis for a calculation model that will reproduce the main features of exit recirculation and will point to its causes.

The appearance of exit recirculation depends on the energy imparted to the fluid on different streamlines at the pump exit, which in turn is intimately related to the exit blade angles (Fig. 14.3). If the blade trailing edge is inclined to the axial direction, the flow on parallel streamlines leaves the impeller at different radii. The circumferential velocity U_2 varies along the trailing edge. The tangential velocity C_{t2} would also have to vary at design conditions if the same energy were to be imparted to the flow on all streamlines. If the meridional velocity C_{m2} remains the same on all streamlines, the exit blade angle β_2 would also have to vary along the trailing edge. The blade angle is measured here from the meridional direction. These relationships derive from the expression of the theoretical head H_{th} of the pump at the design flow rate and apply separately on each streamline:

$$H_{th} = \frac{H_0}{\eta} = \frac{U_2 C_{t2}}{g} = U_2^2 \sigma - U_2 C_{m02} \tan \beta_2$$

where $C_{m02} = Q_0/(\text{impeller exit area})$.

Since $U_2 = \omega r$ varies along the trailing edge and C_{m02} remains the same, for H_{th} to remain constant, the blade angle β_2 must change and compensate for the variation. If H_0 designates the head at the design flow rate Q_0, the relationship between the exit blade angle β_2 and the radius along the trailing edge can be derived from the equations above:

$$\tan \beta_2 = \frac{\omega \sigma r}{C_{m02}} - \frac{H_0}{g \omega \eta r C_{m02}}$$

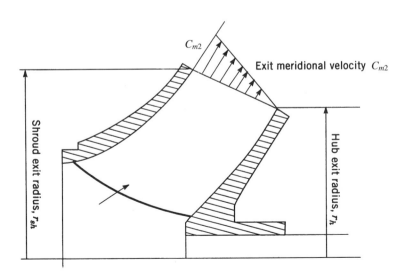

Figure 14.3 Onset of exit recirculation in a mixed flow pump at reduced flow rate.

This theoretical expression prescribes that particular impeller exit blade angle distribution as a function of the radius that will give uniform energy input to the fluid on all streamlines at the impeller exit at the design flow rate. Actual geometrical blade angle distributions may differ from this ideal. However, high efficiency at design flow rate requires that the energy input be uniform. Therefore, the blades of most good pumps will conform approximately to this relationship, or at any rate, produce uniform energy input.

In general, it can be observed that the blade angle β_2 has to become larger (measured from the meridional direction) as the radius becomes greater. In mixed flow pumps the shroud radius is usually larger than the hub radius. Therefore, the exit blades must be more backward swept at the shroud than at the hub. At off-design point operation, when the flow rate Q is reduced, the blade angles being fixed, less energy will be imparted to the flow along the hub than along the shroud, and a shear flow will leave the pump, which is ultimately responsible for exit recirculation. The meridional velocity C_{m2} cannot remain constant along the trailing edge, and eventually reverses at the hub.

To formulate a mathematical expression for the meridional velocity C_{m2} along the trailing edge, the energy equation will first be used to express the local static pressure p_2 at the impeller exit. The energy increase through the pump $\rho U_2 C_{t2}$ is added to the pressure at the inlet p_1, and the kinetic energy of the tangential and meridional velocity components at the trailing edge is subtracted.

$$p_2 = p_1 + \rho U_2 C_{t2} - \rho \frac{C_{t2}^2}{2} - \rho \frac{C_{m2}^2}{2}$$

The radial force balance equation predicts that the static pressure increases radially because of the centrifugal acceleration of the swirling fluid leaving the impeller (Fig. 14.4):

$$\frac{dp_2}{dr} = \rho \frac{C_{t2}^2}{r}$$

The pressure can be eliminated from these two equations by taking the derivative of the expression for the pressure p_2 with respect to the radius and setting it equal to the second equation. This substitution results in a differential equation in terms of the velocities:

$$\frac{d}{dr}\left(U_2 C_{t2} - \frac{C_{t2}^2}{2}\right) - \frac{C_{m2}^2 - C_{t2}^2}{2} \frac{1}{r} = 0$$

Substituting expressions for the circumferential velocity U_2, tangential velocity C_{t2}, and blade angle β_2, an equation is obtained containing only two variables, the radius r and meridional velocity C_{m2}. The equation cannot be solved analytically, in closed form, but a finite difference solution is possible by substituting ratios of

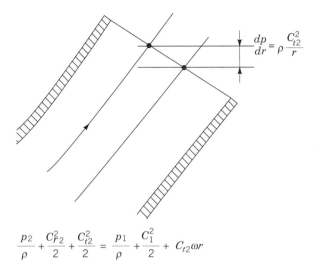

Figure 14.4 Impeller exit flow conditions in a mixed flow pump.

finite increments instead of the derivatives. The trailing-edge length is subdivided into equal increments, and the meridional velocity is calculated at each station.

The reduced flow rate, at which recirculation starts, is obtained by calculating the meridional velocity at several locations along the trailing edge. Flow rate increments are calculated from the product of the meridional velocity and the corresponding, incremental exit cross-sectional areas, which are finally added together to arrive at the total flow rate. Calculation starts at the hub exit radius, where the exit meridional velocity becomes zero when recirculation starts. The velocity is then incremented stepwise along the trailing edge. The computer program listing EXREC for executing the calculation is given in the Appendix.

The flow rate at which recirculation starts, and the corresponding meridional velocity profile, appear to correspond to measured values (Gopalakrishnan et al. 1995; Tuzson 1999). The calculation assumes that the impeller produces uniform angular momentum at the exit at the design flow rate. If the exit conditions of the actual pump differ from the assumed form, the calculations will be in error. The onset of exit recirculation becomes evident from the head–flow curve of the pump. The curve, which approaches the horizontal or even shows a maximum with decreasing flow rate, suddenly starts rising again at the particular flow rate, when recirculation starts as shown in Fig. 14.5. The gradual rise of the head with decreasing flow rate can be explained by the fact that the average effective streamline at which the flow leaves the pump gradually increases with decreasing flow, since recirculation takes increasingly more space near the hub. A greater effective impeller radius generally results in higher head.

An approximate shape of the head–flow curve of a pump can be sufficiently well calculated by considering only an average streamline passing through the

198 INLET AND EXIT RECIRCULATION

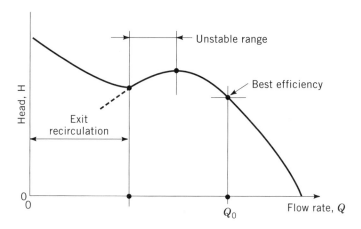

Figure 14.5 Mixed flow pump head–flow curve with exit recirculation.

impeller. In mixed flow impellers the theoretical head usually declines only slowly with increasing flow rate. Under these conditions the actual pump head, which is obtained when the losses are deducted from the theoretical head, often shows a maximum at a reduced flow rate. The operating range below this flow rate, corresponding to the maximum head, is unstable. Indeed, it was shown that a positive slope of the head–flow curve leads to flow oscillations and surging. However, if exit recirculation sets in at or before the maximum is reached with decreasing flow rate, the head starts rising again and a positive slope can be avoided. The onset of exit recirculation does not depend on the conditions on the mean streamline and cannot be calculated with a mean streamline performance prediction program such as the LOSS3 program. Exit recirculation depends on the geometrical configuration of the impeller exit, the slanted trailing edge of the blades, and the blade angle distribution. By tailoring the exit appropriately, the onset of exit recirculation can be controlled, and the pump head–flow characteristics can be adjusted to avoid a positive slope and the corresponding instability.

An important conclusion from the agreement between this calculation procedure and measurements is that exit recirculation results from uneven angular momentum distribution and energy input at the impeller exit. Therefore, the calculation model above has a firm theoretical basis and can be further refined, if desired, with powerful computer programs that can estimate the angular momentum distribution at the impeller exit, or can calculate the exit meridional velocity distribution directly at off-design conditions.

REFERENCES

Batchelor, G. K. (1967): *Fluid Dynamics*, Cambridge University Press, New York, pp. 543–555.

Cooper, P. (1988): Panel Session on Specifying Minimum Flow, *Proceedings of the Texas A&M 5th International Pump User Symposium*, Houston, Texas, pp. 177–192.

Fraser, W. H. (1981): Recirculation in Centrifugal Pumps, *ASME Winter Annual Meeting*, Washington, D.C., November 15–20.

Gopalakrishnan, S. (1988): A New Method for Computing Minimum Flow, *Proceedings of the Texas A&M 5th International Pump User Symposium*, Houston, Texas, pp. 41–47.

Gopalakrishnan, S., Cugal, M., Ferman, R. (1995): Experimental and Theoretical Flow Field Analysis of Mixed Flow Pumps, *2nd International Conference on Pumps and Fans*, Tsinghua University, Beijing, October 17–20.

Graf, E. (1993): Analysis of Centrifugal Impeller BEP and Recirculating Flows: Comparison of Quasi-3D and Navier-Stokes Solutions, in *Pumping Machinery Symposium*, ASME FED Vol. 154, pp. 236–245.

Guelich, J. F., Bolleter, U., Simon, A. (1993): *Feedpump Operation and Design Guidelines*, Sulzer Brothers Ltd., Winterthur, Switzerland. Also McCloskey, EPRI Publication TR-102102, T. Electric Power Research Institute, Palo Alto, Calif., June.

Guesnon, H., Lapray, J. F. (1985): Influence de la géométrie des structures situées à l'aspiration d'une pompe sur les courbes caractéristiques à debit partiel, *La Houille Blanche*, No. 5, pp. 387–404.

Hureau, F., Kermarec, J., Stoffel, B., Weiss, K. (1993): Study of Internal Recirculation in Centrifugal Impellers, *ASME Pumping Machinery Symposium*, FED Vol. 154, pp. 151–157.

Jaberg, H., Hergt, P. (1989): Flow Patterns at Exit of Radial Impellers at Part Load and their Relation to Head Curve Stability, in *Pumping Machinery*, ASME Vol. 81, pp. 213–225.

Jansen, W. (1967): Flow Analysis in Francis Water Turbines, *ASME Journal of Engineering for Power*, July, pp. 445–451.

Kaupert, K. A., Holbein, P., Staubli, T. (1996): A First Analysis of Flow Field Hysterisis in a Pump Impeller, *ASME Journal of Fluids Engineering*, December, pp. 685–691.

Keller, J. J., Egli, W. (1985): Force- and Loss-Free Transition Between Flow States, *Journal of Applied Mathematics and Physics*, ZAMP, Vol. 36, pp. 854–889.

La Houille Blanche (1980): Comportement dynamique des turbomachines hydrauliques, No. 1/2.

La Houille Blanche (1982): Fonctionnement des turbomachines a debit partiel, No. 2/3.

La Houille Blanche (1985): Fonctionnement des turbomachines a regime partiel, No. 5.

Pfleiderer, C. (1961): *Die Kreiselpumpen*, Springer-Verlag, Berlin, pp. 408–416.

Sen, M. (1979): *Inlet Flow in Centrifugal Pumps at Partial Delivery*, Government Publication N81-10437, Von Karman Institute, Rhode Saint Genese, Belgium, June 20–21.

Spannhake, W. (1934): *Centrifugal Pumps, Turbines and Propellers*, MIT Press, Cambridge, Mass., pp. 244–249.

Strscheletzky, M. (1961): Kinetisches Gleichgewicht der Innenströmungen Incompressibler Flüssigkeiten, *VDI-Z Fortschritt Berichte*, Vol. 7, No. 21, pp. 113–132.

Tanaka, T. (1980): *An Experimental Study of Backflow Phenomena in High Specific Speed Propeller Pumps*, ASME Paper 80-FE-1980.

Tuzson, J. (1983): Inlet Recirculation in Centrifugal Pumps, *ASME Symposium on Performance Characteristics of Hydraulic Turbines and Pumps, 1983 Winter Annual Meeting*, Boston, FED Vol. 6, pp. 195–200.

Tuzson, J. (1993): Interpretation of Impeller Flow Calculations, *ASME Journal of Fluids Engineering*, September pp. 463–467.

Tuzson, J. (1999): Impeller Exit Recirculation, Paper FEDFM99-6863, *ASME Fluids Engineering Conference*, San Francisco, Summer.

Vlaming, D. J. (1989): Optimum Inlet Geometry for Minimum NPSH Requirements for Centrifugal Pumps, in *Pumping Machinery*, ASME FED Vol. 81, pp. 25–28.

Von Karman Institute (1978): Off-Design Performance of Pumps, *Lecture Series*, Vol. 3, No. 1, March 6–8.

15
TWO-PHASE FLOW IN PUMPS

TWO-PHASE FLUID BULK PROPERTIES

Although pumps are designed to handle incompressible fluids, at times gases and vapors appear in the liquid and must pass through the pump. In power plant pumps, water near boiling conditions may release water vapor if the pressure of the water is lowered. Process pumps also often handle various organic liquids near boiling and must survive upset conditions by handling liquid–gas or vapor mixtures. Refrigeration typically operates with organic fluids near boiling conditions. Oil well pumps encounter crudes under high pressure, saturated with a variety of dissolved gases, which evolve as the fluid is brought to the surface. Nuclear power plant coolant loss can be feared if because of accidental excess temperatures, the cooling water flashes into vapor. The capability of centrifugal pumps to tolerate and handle liquid–gas mixtures can be an important factor in their application.

Several possible cases can be distinguished regarding the behavior of the two-phase fluid. In one case, gases, which are insoluble in the liquid, may enter the pump with the liquid. In another case, gases originally dissolved in the liquid may come out of solution because of lowered pressures. In still other cases, the liquid itself may vaporize because of lowered pressures. These different cases must be distinguished because the properties characterizing the fluid, in particular the density or specific volume of the fluid, are related differently to the pressure and temperature. The complexity of two-phase flow demands that simplifying assumptions be made to arrive at a manageable mathematical model. The assumptions will differ depending on the particular case.

TWO-PHASE FLOW IN PUMPS

Insoluble gases in the liquid phase or noncondensable gases can be modeled by assuming that the pressure–density relationship of the gas fraction follows the ideal gas law at a constant temperature. Isothermal compression prescribes that the specific volume $v = 1/\rho$ vary in inverse proportion to the absolute pressure p. The constant-temperature assumption remains valid if the gas is present in the form of small bubbles, and any temperature difference between the gas and the liquid can be equalized instantaneously because of the relatively large surface of the bubbles. The assumption breaks down if the gas accumulates in certain locations of the flow passages. Accelerations perpendicular to the streamlines will tend to separate the phases. The rate of separation, the velocity of the bubbles relative to the fluid, will depend on the bubble size. Surface tension presents an additional complicating factor. High surface tension tends to help agglomerate bubbles. Surface energy is reduced if small bubbles with a large aggregate surface accumulate into large bubbles with less total surface area. Small amounts of surface-active agents or even organic impurities, which affect the surface tension, can play an important role. The fluid may appear in the form of a foam, which tends to resist the breakup of bubbles. The following relations define the fluid properties under isothermal conditions:

$$pv = \frac{p}{\rho_g} = RT = \text{constant}$$

$$\rho = \rho_l(1 - \alpha) + \rho_g \alpha$$

In these expressions R is the gas constant, T the absolute temperature in Rankin degrees (Fahrenheit $+ 459.7°$), and ρ_l and ρ_g designate the density of the liquid and the gas. The symbol α stands for the volume fraction of the gas, the void fraction, which itself depends on the gas density or specific volume.

To put things into perspective, a simple calculation would show that uniform spherical bubbles in a rectangular array would touch at a void fraction of $\pi/6 = 0.52$. At a void fraction of 0.2, the width of the liquid layer separating bubbles would amount to about 30% of the bubble diameter. Evidently, coalescence of bubbles could hardly be avoided at void fractions above 0.2, and bubble interactions would be expected to become important well below this limit.

Being compressible, sonic speed may appear in liquid–gas mixtures. The relatively high compressibility of the bubbles and the high inertia of the liquid phase results in a relatively low speed of sound, which can be calculated from the relationship between bulk density and the pressure (Leung 1996). Vaporization of a single-substance liquid introduces thermodynamic effects. Most fluids have a sizable heat of vaporization and cool down when vapors evolve unless heat is supplied from the outside. Consequently, isothermal conditions cannot be taken for granted. Heat transfer between bubbles and the liquid may still be instantaneous, but the heat capacity of the liquid phase must be taken into account.

The case of water mixed with air or steam has great importance in industrial applications and has been investigated intensely. However, care must be taken when experimental results with water are applied to other fluids, because of the exceptional properties of water. Its heat of vaporization, about 1000 Btu/lbm, exceeds the heat of vaporization of other liquids by a factor of 2 or more. Also, water has the highest surface tension among liquids and has the greatest tendency to form large bubbles and to separate the phases.

Two-phase flow conditions in pumps are often compared with similar flows in pipes. It has been recognized that regions of different gas and liquid mixture configurations prevail in pipe flow, depending on the void fraction. Generally, water–air mixtures can be considered to form a homogeneous flowing fluid up to void fractions of about 0.2. Beyond this limit a bubbly flow regime appears, which exists up to a void fraction of about 0.6. At even higher void fractions, separated annular flow has been observed in pipes until the liquid fraction becomes so small that only a fine mist is formed. Starting with the bubbly flow regime, the two phases may move with different velocities. The ratio of velocities, the phase slip, will depend on the interfacial shear force between the phases. These correlations have been based on pipe flow and assume that typically encountered pipe flow turbulent mixing prevails and that no transverse accelerations exits that would tend to separate the phases. In the presence of transverse accelerations, encountered in pumps, the gas and liquid phases would tend to separate, and with the reduction of the interfacial area between the phases, interaction between them would be minimized. The bubbly flow regime could then be expected to disappear in pumps and separated flow may take its place.

In these various flow regimes, different kinds of interactions and drag forces between the phases exist, depending primarily on the degree of dispersion of the gas in the liquid. In the bubbly flow regime, interfacial surface and shear between the phases increases with decreasing bubble size. The bubble size itself depends on the intensity of mixing brought about by the turbulence in the flow, which is difficult to estimate. Analytical studies fall back on empirical correlations, which are valid only in the range for which they have been developed. These drag forces link the motion of the gas and liquid phases and influence the relative velocity between the phases. A difference in velocity between the phases can bring about an accumulation or reduction in the local gas volume, and therefore can affect the local value of the void fraction. In the presence of transverse acceleration, the gas and liquid phases would tend to separate. Obviously, the particular void fraction at which the transition from one regime to the other takes place depends on the physical properties of the liquid. In substances that tend to foam, the transition occurs at higher void fractions (Lea and Bearden 1982).

EXPERIMENTAL PUMP PERFORMANCE DATA

The anticipated possibility of a nuclear reactor coolant flow accident triggered intensive research on the performance of coolant circulating pumps with water and steam or air mixtures. The electric power generation industry supported several full-sized and reduced model tests (Fujie and Yamanouchi 1985; Kamath and Swift 1982; Kim 1983; Mikielewicz et al. 1978; Minato et al. 1985; Runstadler 1976; Wilson 1977; Winks 1977). Detailed data can be found in the original publications. Two-phase pump flow technology was reviewed, data reduction and scaling issues were considered, and empirical correlations were developed (Rundstadler 1976; Wilson 1977). These investigations and their results were summarized in an article by Furuya (1985).

Data were taken at design conditions and were presented in the form of a normalized head and normalized torque. The head was normalized by dividing with the head on pure liquid. The torque was also corrected for density by dividing it with the bulk two-phase density at the pump inlet. These corrected values were plotted against the void fraction as shown in Figs. 15.1 and 15.2 (Furuya 1985; Kamath and Swift 1982).

The normalized head is seen to remain close to 1, from a void fraction of zero, which corresponds to pure liquid, to about 0.2. At higher void fractions it gradually declines to about 0.1 at a void fraction of 0.8. Void fractions beyond 0.8 correspond practically to pure gas and represent conditions in a compressor rather than a pump. The normalized torque also starts from 1 and drops suddenly to

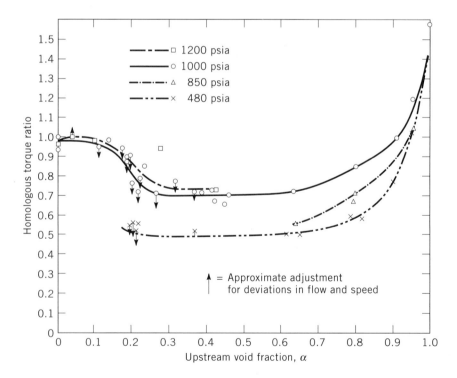

Figure 15.1 Two-phase density-corrected torque ratio of a centrifugal pump as a function of inlet void fraction. (From Kamath and Swift 1982.)

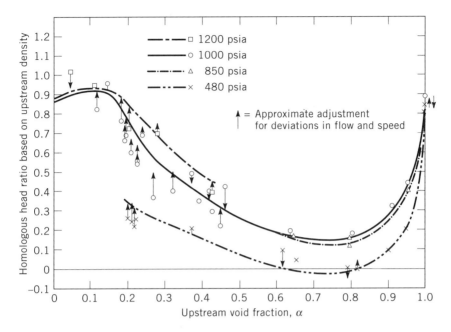

Figure 15.2 Two-phase density-corrected head ratio of a centrifugal pump as a function of inlet void fraction. (From Kamath and Swift 1982.)

between 0.5 and 0.7 at a void fraction of 0.2. From then on it remains approximately constant up to void fraction values even beyond 0.8.

Similar experiments were performed in Japan on behalf of the nuclear power generation industry (Noghrehkar et al. 1995). Analysis results conforming with test data show a sudden drop of normalized head at void fractions of 0.2 or slightly higher. The test results are similar to those in Figs. 15.1 and 15.2. Tests and analysis executed in Germany for an axial nuclear reactor cooling pump attribute the head deterioration to an increase in the impeller slip coefficient resulting from a "displacement of centre of mass due to phase separation" in the impeller (Schneider and Winkler 1988). The relative velocity of the liquid phase moving to the leading face of the blade, in the direction against the sense of rotation, is claimed to diminish the effective tip speed of the impeller. A change in the void fraction at the impeller exit would also contribute to head deterioration. Both of these factors would depend on the inlet void fraction. The experimental head data resemble those of Figs. 15.1 and 15.2.

Entirely different kind of experiments were described by Lea and Bearden (1982). The study intended to address the problem of pumping crude containing dissolved gases from deep oil wells. Tests on water and air, as well as with mixtures of kerosene and CO_2, measured the performance of small vertical pump stages across the operating range. Some of these test data are shown in Fig. 15.3. On air–water mixtures at design conditions and low void fractions up to about 11% air, the head decreases in proportion to the bulk density. At 14% air, the head drops to about 10% of its value on water. On kerosene and carbon dioxide mixtures, which were observed to foam, complete head loss was not reached until void fractions of 40 to 50%. In particular, only about 20% head drop was measured with 50% gas when the inlet pressure was raised to 400 psig, which amounted to 10 times the pressure rise through a single stage. The head tends to decline rapidly with decreasing flow rate below the rated flow, as shown in Fig. 15.3. However, the corresponding positive slope of the pump characteristics induced instability and surging in every instance. The compressibility of the fluid further enhances the instability. At flow rates higher than the design flow rate, choking phenomena appear which are the equivalent of choking in centrifugal

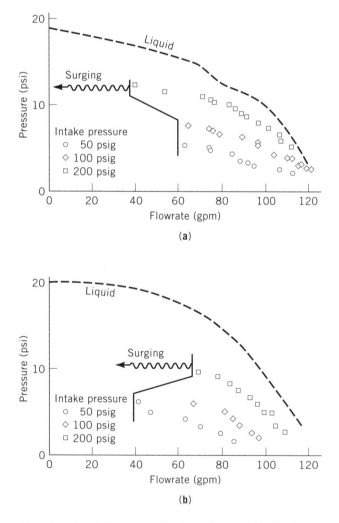

Figure 15.3 Two-phase head–flow curve of centrifugal pump with 15% by volume CO_2 in fuel oil: (a) K-70 pump; (b) C-72 pump. (From Lea and Bearden 1982. Copyright SPE.)

compressors and result from sonic flow at the impeller inlet. At choking the characteristics drop suddenly to zero head and preclude operation at higher flow rates. Indeed, two-phase fluids, being compressible, can experience supersonic flow. The compressibility of the gas bubbles and the high inertia of the liquid phase can combine to yield relatively low values of the speed of sound.

The tests with kerosene and carbon dioxide were performed on three different pump designs: on low-specific-speed radial stages and on high-specific-speed radial and mixed flow stages. The nominal head and flow ratings of the radial and mixed flow stages were the same. Therefore, their specific speeds were also the same. Tests showed that the mixed flow design handled gas–liquid mixtures better. The explanation might be that radial designs rely more on producing a static pressure rise through the impeller, while mixed or axial flow pumps produce an energy increase mostly in the form of a high-velocity head at the impeller exit. With a two-phase fluid it is primarily the static pressure rise through the impeller that is lost. Consequently, high-specific-speed mixed flow stages would be expected to suffer less.

For the eight low-specific-speed radial stages, rated at a nominal 50 gpm and 35 ft per stage, performance curves are published for inlet pressures of 50, 100, and 400 psi and for gas fractions up to 50%. The curves at high inlet pressure show the least deterioration. These performance trends can be explained by the reduction in flow rate in consecutive stages due to the gradual compression of the gas phase. At low inlet pressures the volumetric flow rate in the last stage is reduced to about 70% of the inlet flow rate, assuming isothermal compression of the gas. Solubility of CO_2 in kerosene would be expected to reduce the volumetric flow rate even further. At high inlet pressures the last stage still handles close to 90% of the inlet flow rate. With reduced flow velocities, separation of the phases would proceed faster. Satisfactory pump operation with such surprisingly high void fractions is possible only because CO_2 in kerosene tends to foam and resists phase separation better than do water and air or steam.

Important tests have also been performed measuring the pressure recovery in diffusers with two-phase flow (Hench and Johnston 1971). The general trend of the pressure recovery coefficient follows the trend of liquid diffusers and appears closer to a sudden expansion with a maximum near an area ratio of 2, than an ideal diffuser. Little performance deterioration was found up to void fractions of 0.38, but pressure recovery dropped at higher void fractions and vanished at a void fraction of approximately 0.7. Pressure kept increasing beyond the diffuser exit, in the straight pipe section following the experimental diffuser, indicating that considerable slip between the phases might still have been present at the diffuser exit, and that in the following straight pipe section, the faster-moving liquid compressed the gas fraction in the process of equalizing their velocities.

CALCULATION MODELS

Several analytical and empirical correlation methods have been developed, especially on behalf of the nuclear power industry, to represent mathematically the experimental pump data taken with water–air or steam mixtures (Kim 1983). In most instances, a factor, a head multiplier correlation, was developed, the expectation being that the head on water could be multiplied by the multiplier to calculate the two-phase flow head (Kamath and Swift 1982). The multiplier was expected to depend only on the void fraction. In a similar formulation, a head loss ratio correlation was proposed instead (Mikielewicz et al. 1978).

An unspoken assumption in the case of some of these correlations is that the flow pattern with gas–liquid mixtures remains the same as with pure liquid: that "the trajectory of the liquid be identical to that of bubbles" (Furuya 1985). Such may be the case for low void fractions and small bubbles (Minemura and Murakami 1980). However, beyond a void fraction of about 0.2, the phases separate and an entirely different flow pattern appears in the impeller which bears no relationship to the flow pattern on liquid water. Therefore, an entirely different model is required for higher void fractions, which cannot be grafted onto the liquid flow pattern.

The simplest analytical model assumes that pump performance remains the same with the two-phase fluid, except for a change in bulk density. The assumption implies that the pressure rise through the pump remains insufficient to produce a significant volume change in the gas phase while passing through the pump. Such is the case when the pump pressure rise is small compared with the absolute pressure of the fluid. A further condition requires that the gas be well dispersed in the fluid and that no significant separation of the phases should take place in the pump. A low void fraction would probably satisfy such a requirement. Validity of the model can be evaluated by calculating the gas volume change from pump inlet to pump exit and estimating the percentage velocity change from the changed volumetric flow at the pump exit.

At low void fractions and in the case of vapors that can condense or gases that can dissolve in the liquid, the possibility exists that the pressure rise through the impeller suffices to significantly reduce or even eliminate the gas volume (Noghrehkar et al. 1995). Even in noncondensable gases, a change in the void fraction from impeller inlet to exit cannot be excluded. Therefore, a further correction to the foregoing model might be to calculate the bulk density change, or the gas volume change, stepwise along the streamline through the impeller from the local static pressure, assuming isothermal conditions.

More elaborate models calculate the flow of gas and liquid phases separately. Accelerations in the direction of the streamlines are taken into account and a slip, a velocity difference between the phases, is stipulated, which can result in a gas holdup, a local increase in the void fraction (Furuya 1985; Grison and Lauro 1979; Mikielewicz et al. 1978; Minato et al. 1985; Murakami and Minemura 1983; Noghrehkar et al. 1995; Zakem 1980, 1987). The velocity difference between the phases implies mutual interaction between phases. Additional assumptions need to be introduced regarding the drag forces between phases, which in turn depend on the bubble sizes. Since theoretical bubble dynamics do not allow generally valid predictions, empirical data are introduced which derive from test conditions similar to those of the intended application, but may lack general validity. Bubble size, gas dispersion, and mixing in the liquid should be a function of the turbulence. Specialized turbulence models exist for two-phase flow which could be linked to the bubble motions (Lopez de Bertodano et al. 1994).

The test data in Fig. 15.1 lead to the important observation that beyond a void fraction of 0.2, the density-corrected torque does not depend on the void fraction. This fact implies that the theoretical head, the energy imparted to a unit volume of fluid by the impeller, remains the same regardless of the void fraction. Once the flow separates, the void fraction at the impeller exit is irrelevant. Therefore, the variation of the head with the void fraction shown in Fig. 15.2 cannot result from a variation of the theoretical head but must result from head losses. Such head losses can be present at the inlet or in the volute and diffuser. Since no significant static pressure rise can take place in the impeller in the presence of gas, the energy input must be present in the form of velocity head at the impeller exit. Major head losses occur when velocity head is transformed into pressure head by the diffuser. Diffuser losses probably account for the head losses in two-phase pumps. The diffuser can consist of a high-speed jet exhausting into a straight flow passage since even a sudden expansion can recover up to 50% of the jet's kinetic energy.

FULLY SEPARATED FLOW MODEL

A drastic simplification consists of assuming that the liquid and gas completely separate in the impeller. A liquid layer would then be present on the side of the blades facing the direction of rotation (Fig. 15.4). The gas would collect on the suction side. Since the gas region would extend from the inlet to the exit and could support practically no radial pressure increase, the energy imparted to the fluid by the pump would consist exclusively of the kinetic energy of the liquid at

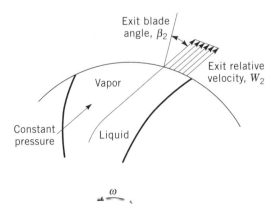

Figure 15.4 Model of fully separated two-phase flow in a pump impeller.

the pump exit. Applying the rothalpy I equation to a streamline in the liquid layer at constant pressure ($p_2 = p_1$), the exit relative velocity would become

$$\frac{P_1}{\rho} + \frac{C_1^2}{2} = \frac{p_1}{\rho} + \frac{W_1^2 - U_1^2}{2} = I = \frac{p_1}{\rho} + \frac{W_2^2}{2} - \frac{U_2^2}{2}$$

$$W_2 = \left(U_2^2 + C_1^2\right)^{1/2}$$

The principle of conservation of angular momentum still applies and can be used to calculate the absolute tangential velocity C_{t2} at the impeller exit, noting that no slip takes place since the flow in the liquid layer follows the blade.

$$C_{t2} = U_2 - kW_2 \sin \beta_2$$

$$H_{\text{th}} = \frac{U_2 C_{t2}}{g}$$

Here H_{th} is the theoretical head, the energy imparted to the liquid per unit volume, and β_2 is the blade angle measured from the radial in the direction opposite to the sense of rotation. The symbol k stands for a coefficient reducing the relative velocity W in the impeller as a consequence of losses.

The shaft torque T can be calculated from the product of liquid density ρ_l, the theoretical head H_{th}, and the liquid flow rate $Q(1 - \alpha)$, by dividing their product by the shaft speed N:

$$T = \frac{\rho_l H_{\text{th}} Q(1 - \alpha)}{N}$$

It will be noted that when fully separated flow exists, the bulk density-corrected torque $T/(1 - \alpha)$ will not depend on the void fraction α and will remain constant since the theoretical head remains constant. Consequently, the power input to the pump will be in proportion of the liquid flow rate into the pump.

It is also apparent that the bulk flow rate affects the relative velocity at the exit W_2 only slightly. Only the inlet velocity C_1 depends on the flow rate. It is usually much smaller than the tip speed U_2:

$$W_2 = \left(U_2^2 + C_1^2\right)^{1/2}$$

Consequently, when the phases separate in the impeller, the theoretical head will not depend on the void fraction and will depend only slightly on the flow rate. This mode of operation differs significantly from that with liquid, in which case the theoretical head increases linearly with decreasing flow rate. One consequence of this change in flow pattern is that the slope of the pump characteristics decreases and a maximum of the head appears at a reduced flow rate. Pump operation below this flow rate becomes unstable since the head–flow characteristic has a positive slope.

To calculate the net head H, the theoretical head H_{th} would need to be multiplied by the efficiency η. As a first approximation it could be assumed that the losses result entirely from diffusion losses, and that therefore the pump efficiency numerically equals the diffuser pressure recovery coefficient, which is available from test results (Hench and Johnston 1971). The liquid sheets leaving each blade at the impeller exit appear for a stationary observer at the impeller exit to form alternating liquid and gas layers at the inlet to the diffuser vanes or the volute tongue. They will strongly mix and produce a homogeneous two-phase fluid at the entrance of the diffuser section. Therefore, well-mixed fluid conditions exist at the diffuser inlet, and flow separation is unlikely at that location. A good diffuser design may act as a two-phase ejector, which entrains gas with a high-velocity liquid jet. Efficiencies would then vary and can be approximately correlated for void fractions beyond 0.2 by the empirical expression which approximately follows the experimentally found pressure recovery coefficient in a two-phase diffuser (Hench and Johnston 1971):

$$C_p = \eta = 1 - 1.25\alpha$$
$$H = H_{th}\eta$$

Consequently, the head of the pump on two-phase flow will suddenly drop near a void fraction of about 0.2, to a level of 0.5 to 0.7, because of the drop in the theoretical head, and decay with void fraction toward zero at very high values because of a declining pressure recovery coefficient, which is numerically equal to the pump efficiency.

At off-design conditions the inlet flow contacts the leading edge of the blades, and flow separation would be expected to occur. Some of the liquid will spread to the hub and shroud surfaces and may not accumulate on the blade leading face. The possibility of inlet recirculation may exist. With a separation of the liquid and gas phases at the impeller inlet, the pump may lose its prime and stop pumping, or enter an irregular surging mode of operation. Without liquid in the impeller, the pump cannot aspirate fluid into the inlet. The appearance of a maximum on the head–flow characteristic of the pump will contribute to the instability of the pumping system. All these phenomena have been observed in pump tests with liquid–gas mixtures. The specific conditions when these flow pattern appear depend on the particular design of the pump.

COMPARISON WITH TEST DATA

The fully separated two-phase pump flow model contains two empirical quantities: the dimensionless loss coefficient k and the efficiency or diffuser pressure recovery coefficient $\eta = C_p$. The loss coefficients can be determined from the extensive tests performed on the nuclear reactor cooling pumps, in particular from Figs. 15.1. and 15.2. The specific test data used here come from the one-fifth scale model of the mixed flow pump, having the specifications 252 ft (77 m) at 3500 gpm (0.221 m³/s) and 4500 rpm, corresponding to a specific speed of 4200. The model had a 6-in. (0.1525-m)-diameter inlet flange and a nominal impeller diameter of 7.75 in. (0.2 m). The exit blade angle was 65°. Other geometrical dimensions can be scaled from the cross-sectional drawing of the pump (Winks 1977, Fig. 4–8). A performance calculation performed with these data will approximately reproduce the published head–flow curve of the pump. The inlet velocity becomes $C_1 = 40$ ft/sec (12.2 m/s), the tip speed $U_2 = 152$ ft/sec (46.4 m/s), and the impeller exit relative velocity $W_2 = 157$ ft/sec (47.8 m/s). The experimental data show a density-corrected torque value that is 0.5 to 0.7 times the value on liquid. It remains practically constant from a void fraction of 0.2 to very high values. Analytical considerations show that at constant rotational speed, torque is proportional to the absolute tangential velocity at the impeller exit, which in this instance, for pure liquid, takes on the value of $C_{t2} = 53.4$ ft/sec (16.3 m/s). On liquid–gas mixtures it would be 0.5 to 0.7 times as much. These test results will agree with the value calculated from the fully separated model if the loss coefficient were to take on the value of $k = 0.88$ to 0.8. These values correspond to a velocity head loss in the impeller amounting to 25 to 35% of the velocity head $W_2^2/2g$, which does not appear unreasonable.

The theoretical head H_{th}, proportional to the absolute tangential velocity at the impeller exit C_{t2}, remains independent of the void fraction. Indeed, the separated liquid layer on the blades leaves the impeller with the same velocity regardless of the void fraction, which affects only the layer thickness, not its velocity. The gradual decline of the actual head results entirely from diffuser losses, which also account for the drop in efficiency. Using the two-phase diffuser pressure recovery data of Hench and Johnston (1971), good agreement can be obtained with the test data. For void fractions from 0.2 to 0.6, a linear relationship appears to be appropriate between the pressure recovery coefficient C_p, or the efficiency η, and the void fraction α:

$$C_p = \eta = 1 - 1.25\alpha$$

Considering the drastic simplifications, the test data agree satisfactorily with the calculation. The real flow pattern in the impeller is certainly much more complicated than the fully separated flow model would lead us to believe. After all, it is only an analytical model. However, it illustrates the extreme case and can bracket the actual state in the impeller. The separation of gas bubbles from a homogeneous mixture can certainly not go beyond total separation. This simple model illustrates and explains, at least qualitatively, the peculiar phenomena that appear in centrifugal pumps when operating on liquid and gas mixtures.

FILL–SPILL FLUID COUPLING

The application of the fully separated two-phase flow model to a fill–spill fluid coupling is a most appropriate example (Uchiyama et al. 1981). A fluid coupling

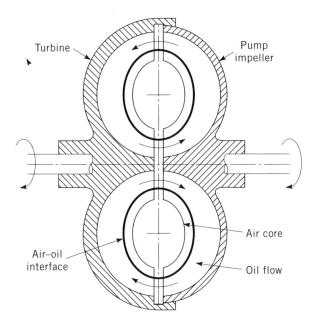

Figure 15.5 Fill–spill fluid coupling.

consists of a close-coupled centrifugal pump and centripetal turbine with a common axis of rotation as shown on Fig. 15.5. Power is supplied to the pump and is transmitted to the turbine by fluid forces. Depending on the magnitude of the input torque and the capacity of the coupling, some slip, a velocity difference between the input and output shafts, exists. Since the same torque must act on the pump as on the turbine, the slip corresponds to the power transmission inefficiency or loss. Torque converters operate in a similar fashion, but torque amplification or reduction becomes possible only if a stationary reaction element, a bladed stator, exists somewhere in the fluid circuit (Brun and Flack 1996; Föster 1960; Lucas and Rayner 1970).

The flow takes on a toroidal shape and cycles through the pump and turbine repeatedly. The power transmitted by the coupling from the pump to the turbine corresponds to the angular momentum produced by the pump, or the head, multiplied by the mass flow. Power transmission efficiencies above 90% can be achieved since the flow never slows down, and no diffusion losses exist. These high efficiencies prove that the inefficiency of centrifugal pumps results mostly from diffusion losses.

In fill–spill couplings, the torque capacity and the slip, the velocity difference between the input and output shafts, is adjusted by admitting air to the center of the toroidal flow. Because of the centrifugal acceleration the air remains separated from the liquid in the center of the torus. The net flow cross section and, as a consequence, the flow rate are reduced by the presence of the air, reducing the torque capacity of the coupling in this manner.

It is to be noted that the interface between the air core and the liquid remains at constant pressure. The pump adds energy to the fluid by increasing its kinetic energy. No pressure rise can take place. The turbine is of the pure impulse type, converting the kinetic energy of the fluid to shaft power. The flow pattern in the fill–spill coupling shows that the presence of gas–liquid mixtures in a pump impeller does not reduce the impeller's capability to impart energy to the fluid efficiently, that most energy losses result from diffusion, and that the flow pattern takes on an entirely different configuration than in an impeller filled with liquid alone.

REFERENCES

Brun, K., Flack, R. D. (1996): *The Flow Field Inside an Automative Torque Converter: Laser Velocimeter Measurements*, SAE Technical Paper 960721.

Förster, H. J. (1960): Föttinger-Wandler und-Kupplungen für Kraftfahrzeuge, *Automobil Industrie*, April, pp. 53–83.

Fujie, H., Yamanouchi, A. (1985): A Study on Applicability of Similarity Rule to Performances of Centrifugal Pumps Driven in Two-Phase Flow, *Nuclear Engineering Design*, Vol. 85, pp. 345–352.

Furuya, O. (1985): An Analytical Model for the Prediction of Two-Phase (Noncondensable) Flow Pump Performance, *ASME Journal of Fluids Engineering*, Vol. 107, March, pp. 139–145.

Grison, P., Lauro, J. (1979): Comportement des pompes en régime diphasique, *La Houille Blanche*, No. 5, pp. 405–411.

Hench, J. E., Johnston, J. P. (1971): *Two-Dimensional Diffuser Performance with Subsonic, Two-Phase, Air–Water Flow*, ASME Paper 71-FE-20.

Kamath, P. S., Swift, W. L. (1982): *Two-Phase Performance of Scale Models of a Primary Coolant Pump*, EPRI Paper NP-2578, September.

Kim, J. H. (1983): Perspective on Two-Phase Flow Modeling for Nuclear Reactor Safety Analysis, *ASME Cavitation and Multiphase Forum*.

Lea, J. F., Bearden, J. L. (1982): Effect of Gaseous Fluids on Submersible Pump Performance, *Journal of Petroleum Technology*, December, pp. 2922–2930.

Leung, J. C. (1996): On the Application of the Method of Landau and Lifshitz to Sonic Velocities in Homogeneous Two-Phase Mixtures, *ASME Journal of Fluids Engineering*, March. pp. 186–188.

Lopez de Bertodano, M., Lahey, H. T., Jr., Jones, O. C. (1994): Development of a k–ε Model for Bubbly Two-Phase Flow, *ASME Journal of Fluids Engineering*, March, pp. 128–134.

Lucas, G. G., Rayner, A. (1970): Torque Converter Design Calculations, *Automobile Engineer*, February, pp. 56–60.

Mikielewicz, J., Chan, T. C., Wilson, D. G., Goldfinch, A. L. (1978): A Method for Correlating the Characteristics of Centrifugal Pumps in Two-Phase Flow, *ASME Journal of Fluids Engineering*, December, p. 395.

Minato, A., Yamanouchi, A., Narabayashi, T. (1985): Estimation of Centrifugal Pump Head in Steam-Water Two-Phase Flow, *Journal of Nuclear Science and Technology*, May, pp. 379–386.

Minemura, K., Murakami, M. (1980): A Theoretical Study on Air Bubble Motion in a Centrifugal Pump Impeller, *ASME Journal of Fluids Engineering*, Vol. 102, December, pp. 446–455.

Murakami, M., Minemura, K. (1983): Effects of Entrained Air on the Performance of Horizontal Axial-Flow Pumps, *ASME Journal of Fluids Engineering*, Vol. 105, December, pp. 382–388.

Noghrehkar, G. R., Kawaji, M., Chan, A. M. C., Nakamura, H., Kukita, Y. (1995): Investigation of Centrifugal Pump Performance Under Two-Phase Flow Conditions, *ASME Journal of Fluids Engineering*, March, pp. 129–137.

Runstadler, P. W., Jr. (1976): *Review and Analysis of State-of-the-Art of Multiphase Pump Technology*, EPRI Paper NP-159, February.

Schneider, K., Winkler, F. J. (1988): Physical Model for Reactor Coolant Pumps, *Nuclear Engineering and Design*, Vol. 108, pp. 99–105.

Uchiyama, K., Takagi, T., et al. (1981): Rotational Speed Fluctuation and Internal Flow in a Variable-Filling Fluid Coupling, *Bulletin of the JSME*, Vol. 24, January, pp. 109–116.

Wilson, D. G. (1977): *Analytical Models and Experimental Studies of Centrifugal Pump Performance in Two-Phase Flow*, EPRI Paper NP-170, January.

Winks, R. W. (1977): *One-Third-Scale Air-Water Pump Program, Test Program and Pump Performance*, EPRI Paper NP-135, July.

Zakem, S. (1980): Determination of Gas Accumulation and Two-Phase Slip Velocity Ratio in a Rotating Impeller, *ASME Polyphase Flow and Transport Technology Symposium*, pp. 167–173.

Zakem, S. (1987): *Correlation for Two-Phase Flow Pump Performance*, ASME Paper 87-FE-5.

16

HIGH-VISCOSITY PUMPS

PERFORMANCE ESTIMATES AT HIGH VISCOSITY

Stepanoff offers correction factors to convert pump performance data from tests on water to performance on viscous fluids in particular hydrocarbons. Since the correction factors have been derived from test data they remain valid and can give satisfactory results (Stepanoff 1957).

More recently, the Hydraulic Institute published graphical procedures for estimating the effect of viscosity on the head, flow rate, and efficiency. Correction factors for head, flow rate, and power are presented as functions of head, flow rate, and viscosity. Higher viscosities than the viscosity of water are encountered, for example, in hydrocarbon fluids. Standard testing procedures (ISO, API, ASME) also contain accepted methods for correcting test data for the effect of viscosity (Lünzmann and Kosyna 1993). Test data have been published by (Hergt et al. 1981), and recently, calculation procedures have been given by Guelich (1999). The pump performance calculation computer program LOSS3, presented in this book, offers an easy approach, which requires only that some of the loss coefficients be adjusted for high viscosity (Tuzson and Iseppon 1997).

A variety of measurement units are in use to express the numerical value of viscosity. Dynamic μ and kinematic $\nu = \mu/\rho$ viscosity must be distinguished, ρ being the density of the fluid. At standard ambient conditions the dynamic viscosity of water is 21.1×10^{-6} lbf sec/ft^2 (1×10^{-3} N s/m^2) and the kinematic viscosity is 10.9×10^{-6} ft^2/sec (1×10^{-6} m^2/s), the density being 1.939 lbf sec^2/ft^4. Metric units use the centipoise (cP; 1/1000 N·s/m^2) to express the dynamic viscosity, which is just about its value for water. The viscosity of lubricating oils, for example, can be several hundreds of times that of water and depends very strongly on the temperature.

Viscosity affects only certain type of losses in the pump—skin friction, disk friction, and diffusion losses—designated as SKIN, DISK, and CVD in the performance calculation computer program. The dimensionless coefficients, corresponding to these losses, depend on the Reynolds number, as shown schematically in Fig. 16.1. The Reynolds number is the product of the pertinent flow velocity and the typical geometrical dimension divided by the kinematic viscosity, $Re = VL/\nu$. In each case the coefficient remains approximately constant at high Reynolds numbers such as those encountered in water pumps. With increasing viscosity the Reynolds number decreases and reaches a transition range at a certain value. Below the transition range the loss coefficients depend in inverse proportion on the Reynolds number and therefore, if the geometry and flow velocity remain the same, increase in direct proportion to the viscosity.

Mathematical expressions and plots for determining the numerical value of pipe friction or wall friction coefficients have been presented in Chapter 4. The Reynolds number dependence of loss coefficients and pump efficiency is discussed in great detail in an article by Osterwalder and Hippe (1982). In the case of the skin friction loss (SKIN), the average through-flow velocity and the flow passage diameter, or hydraulic diameter, represent the characteristic variables entering into the Reynolds number. The skin friction coefficient λ is available from handbooks and publications. Empirical correlations have been given in earlier chapters of the book. The plot follows the general trend described above. The numerical value of the coefficient at high Reynolds numbers, and the Reynolds number at which the transition occurs, depend on the relative roughness of the passage walls. The transition Reynolds number varies between 2000 and 10,000. Depending on the Reynolds number, the friction coefficient in the viscous and turbulent ranges is given by

$$\lambda = \frac{64}{Re}$$

$$\lambda = \frac{1}{[2\log(0.5D/k) + 1.74]^2}$$

In this expression k is the average roughness height as compared with the flow passage size characterized by the diameter D (Schlichting 1960).

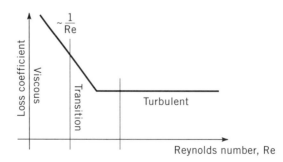

Figure 16.1 Logarithm of the friction coefficient versus logarithm of the Reynolds number.

The coefficient C_m of the disk friction loss DISK follows a similar trend (Nece and Daily 1960). In this case the Reynolds number is based on the impeller tip speed U_2 and the impeller diameter D_2. However, the side clearance space, between the impeller or disk and the housing must be larger than the boundary layer that could be expected at the viscosity and rotational velocity under consideration, as discussed below. The coefficient remains approximately constant at high Reynolds numbers, beyond the transition range around $Re = 10^4$ to 10^5 and has a value from 0.006 to 0.01. The viscous range can be approximated by

$$C_m = \frac{300}{Re}$$

In the performance calculation computer program the mechanical losses of the pump, which result primarily from bearing friction, are lumped together with the disk friction losses. Sleeve bearings already operate in the viscous flow regime. Therefore, bearing losses will increase in direct proportion to the viscosity if sleeve bearings are used and if the fluid passing through the pump lubricates the bearings. If rolling element bearings are used or if the bearings are separately lubricated with lubrication oil, they obviously remain unaffected by the viscosity of the fluid being pumped.

The coefficient of the diffusion losses (CVD) would also be expected to depend on the viscosity. When viscous forces predominate, the kinetic energy of the fluid can be neglected. Not much, if any, of the velocity head of the fluid can be transformed into pressure head. In this case the Reynolds number should be based on the flow passage hydraulic diameter and the flow velocity. At high Reynolds numbers, when inertia forces predominate, the coefficient remains constant. As in pipe flow, the transition could be expected at a Reynolds number of around 10^3, below which the loss coefficient would increase in inverse proportion to the Reynolds number. Although tabulated values or a graphical plot showing the loss coefficient against the Reynolds number are not available, test data on water can be extrapolated to high-viscosity fluids by assuming that the experimental coefficient would begin to rise approximately at the transition Reynolds number. Proposed loss coefficients to be used in the performance calculation computer program LOSS3 (Tuzson and Iseppon 1997) are shown in Table 16.1. Computer calculation results shown in Fig. 16.2 using loss coefficients adjusted according to the guidelines above have been compared with test data and support the approach suggested for estimating the effect of viscosity on pump performance. The experimental data presented by Stepanoff (1993) show a similar trend.

DESIGN FOR HIGH VISCOSITY

The insignificance of inertia forces, compared with viscous forces, governs pump design for high viscosity. The head delivered by the pump approximately equals

Table 16.1 Loss coefficients at high viscosities

| | \multicolumn{5}{c}{Viscosity (cP)} | | | | |
	1	50	100	200	500
SKIN	0.008	0.035	0.06	0.1	0.2
DISK	0.005	0.02	0.04	0.08	0.2
CVD	0.9	0.9	0.9	1.8	4.5

the static pressure produced at the impeller exit since the kinetic energy of the flow at the impeller exit cannot be recovered. Diffusers are ineffective. It is found that all other things remaining the same, the smaller the tangential velocity C_{t2} at the impeller exit, the less head the pump will produce but the greater will be the fraction in the form of static head. Only efficiencies below about 50% can be expected. Therefore, ideally, viscous pumps should be designed for low tangential velocity at the impeller exit C_{t2}; consequently, for low head coefficients, $\varphi = C_{t2}/U_2$, which implies strongly backward-curved blades, corresponding to large exit blade angles β_2.

The major losses in viscous pumps will come from the close clearances that exist in sleeve bearings and wear rings. These close clearances affect pump performance by the frictional torque they produce and by the leakage flow through the clearance. The bearing torque T will increase in direct proportion to the dynamic viscosity μ, and the angular velocity ω and the length L in proportion to the square of the diameter D and in inverse proportion to the clearance width c, as shown below. The leakage rate q will increase with the diameter D, the pressure difference $(p_2 - p_1)$, and the cube of the clearance width c, but will decrease in direct proportion to the dynamic viscosity μ and

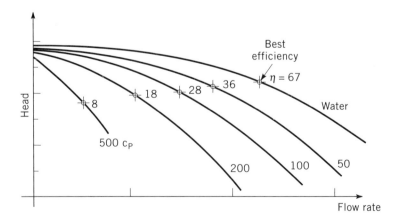

Figure 16.2 Calculated head–flow curve of a centrifugal pump on high-viscosity fluids.

length L. If the torque and leakage are to be minimized at the same time, the effect of increased viscosity should be compensated by decreasing the length. The length of sleeve bearings must be reduced only in proportion to the viscosity increase, since the supporting bearing forces must be maintained. The clearance can also be opened to reduce the torque, but this modification will increase the leakage rate disproportionately.

$$T = \frac{\mu\omega\pi D^2 L}{2c}$$

$$q = \frac{\pi D c^3 (p_2 - p_1)}{24\mu L}$$

In high-viscosity pumps, attention must also be paid to the clearance spaces between the impeller and the stationary housing. If the boundary layer on the rotating face of the impeller is larger than the width of the clearance space, the disk friction torque will depend on the clearance width and will generally be greater than otherwise. An approximate estimate of the boundary layer thickness is a low multiple of $(v/\omega)^{1/2}$ (Schlichting 1960, pp. 86, 178). Consequently, the side clearance in high-viscosity pumps may have to be opened up. Changes in the direction and magnitude of the flow velocity can be better tolerated in pumps for high viscosity, since the inertia and the velocity head of the flow do not matter.

REFERENCES

Guelich, J. F. (1999): Pumping Highly Viscous Fluids with Centrifugal Pumps, *World of Pumps*, August/September, pp. 30–34.

Hergt, P., et al. (1981): Verlustanalyse an einer Kreiselpumpe auf des Basis von Messungen bei hoher Viscosität des Fördermediums, *VDI Berichte*, No. 242.

Lünzmann, H., Kosyna, G. (1993): Erforschung der Verfahrensparameter und Ermittlung von Korrecturfactoren für Pipeline-Kreiselpumpen beim Fördern von Mineralien mit Erhöter Viscosität, *DGMK Forschungsberichte*, No. 479, February.

Nece, R. E., Daily J. W. (1960): Roughness Effects on Frictional Resistance of Enclosed Disks, *ASME Journal of Basic Engineering*, September, pp. 553–562.

Osterwalder, J., Hippe, L. (1982): Studies on Efficiency Scaling Process of Series Pumps, *IAHR Journal of Hydraulic Research*, Vol. 20, No. 2, pp. 175–199.

Schlichting, H. (1960): *Boundary Layer Theory*, McGraw-Hill, New York, pp. 86, 178, 517.

Stepanoff, A. J. (1957): *Centrifugal and Axial Flow Pumps*, Wiley, New York. Reprinted by Krieger Publishing, Malabar, Flo., 1993.

17

SAND AND SLURRY EROSION IN PUMPS

APPLICATIONS AND REQUIREMENTS

Erosion from sand particles might be accidental in pumps handling surface water, such as irrigation, dewatering or flood control pumps, pump-turbines in hydropower installations, or pumps in wells handling fluids with suspended sand particles. In these applications low solid particle concentrations are encountered. Erosion damage accumulates over a considerable time period. One might also assume that the particles are small, do not settle out easily, and remain suspended (Addie et al. 1996; Truscott 1972; Wellinger and Uetz 1955).

Dredging pumps must handle high concentrations of solids, sometimes consisting of large pebbles. However, the particles, consisting primarily of beach sand, are well rounded and less abrasive than freshly crushed material. Pumps applied in coal or ore beneficiation processes encounter severe specialized requirements. Ore is crushed and wet-ground down to small size in ball or rod mills, in order to produce separate ore and rock particles. The outlet from the mills is fed to hydrocyclones by large slurry pumps. These centrifugal separators pass on the particles, which are fine enough, to the ore separation process, while returning large particles to the mills. The slurry pumps used in these grinding circuits handle high-concentration slurries consisting of freshly crushed, relatively fine slurry particles, which are very abrasive. The question is not whether the pumps will wear out, but when. Pumps are sometimes rated by the total tonnage of ore they can process before they wear out.

Since erosion from slurries accumulates progressively, calculations must estimate the rate of material removal. In most slurry pump applications the monetary loss due to pump failure consists of lost production rather than the cost of the pump, since the process must be shut down when the pump fails. If the erosion rate can be estimated, the pump can be changed out in time, during scheduled maintenance, instead during an emergency. A second purpose of slurry erosion calculations could be to determine the distribution of the erosion rate within the pump. Ideally, material should be removed by erosion uniformly from all flow passage walls. An excessive localized wear would result in an undesirable premature pump failure.

SLURRY EROSION MECHANISMS

In analogy to sand blasting, the dominant slurry erosion mechanism consists of impact erosion (Fig.17.1). Particles generally move with the fluid but deviate from the streamlines when they encounter accelerations or decelerations transverse to the streamlines. The particles are heavier than the fluid and experience greater inertia forces during acceleration or deceleration. When relatively large particles, in a liquid jet at low concentration, are directed against a solid wall, they do not follow the fluid and turn away but strike the wall. Their kinetic energy is dissipated in the destruction of the wall surface and in material removal (Peterson and Winer 1980). The erosion mechanism has been studied in the process of abrasive machining (Hashish 1984), where a high-velocity water jet containing abrasive particles is used to cut metal. A complex combined cutting and shallow-angle impact mechanism was stipulated.

Experiments under idealized conditions with single spherical balls striking comparatively flat solid surfaces at different angles have determined the theoretical energy dissipation and the remaining energy of the particles rebounding from the wall, the restitution energy (Brach 1984, 1988, 1991). However, particles below a certain size and kinetic energy, immersed in a liquid, do not strike the solid surface but approach it asymptotically (Zenit et al. 1999). Experimental erosion rate data, particles colliding with flat surfaces, have also been correlated and provide an empirical framework for ranking materials of construction (Clark and Wong 1995).

Complications arise at high concentrations and small particle sizes. Most analytical studies assume relatively low concentrations of relatively large

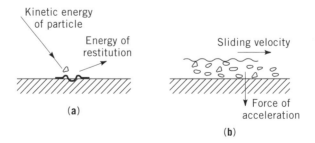

Figure 17.1 (a) Impact and (b) sliding erosion mechanisms from slurry particles.

particles. These assumptions allow that interactions between particles be neglected, which facilitates the calculations. Particle concentration reaches a limit at about 75% by volume, a void fraction of 0.25, when maximum packing occurs; the particles touch each other, and the fluid is contained in the interstices between the particles (Suzuki et al. 1981). Such beds of closely packed particles accumulate near flow passage walls when the acceleration is directed toward the wall. Under such conditions, particles impact the particle bed and cannot reach the wall. Also, when particles become small, on the order of 100 μm (0.1 mm or 4/1000 in.), in general the particle drag forces in water exceed inertia forces, and the particles slow down considerably before reaching the wall (Zenit et al. 1999). Particle inertia decreases with the cube of the particle size, while particle drag only drops off with the square of the particle size.

Therefore, at high concentrations and small particles, a different erosion mechanism must be assumed. If in the extreme, a thick particle bed would become entirely solid, it would be pressed against the wall by its inertia, and would act as sand paper (Mulhearn and Samuels 1962). The erosion mechanism would be scratching and plowing by individual particles (Fig. 17.1). Such an erosion could be characterized as sliding bed erosion (Chen et al. 1980). Between the two extremes of impact erosion and sliding erosion, a transition will exist; when the particle bed is not entirely solid, the particles can slide, roll, and tumble but are still pressed against the wall, which they indent and plow. The pressure force against the wall will depend on the thickness of the particle bed and on the acceleration against the wall.

Whichever erosion mechanism acts, the material removal rate has been found to be approximately proportional to the local rate of power dissipation (Beckmann 1980; Clark and Wong 1995; Maan and Broese Van Groenou 1977; Tuzson 1984). The erosion depth would then be proportional to the local energy dissipated by the slurry particles. Naturally, the material removal rate will depend on the slurry material and on the material of the flow passage wall. The energy required to remove a unit volume of wall material, called the specific energy of the particular slurry and wall material, has to be determined by experiment. Generally speaking, the process whereby the molecular bonds of materials are broken, not only in erosion but also in grinding or crushing processes, is extremely complex and has given rise to several theories, since the corresponding energy requirement is of great interest (Beke 1964). Alternative correlations for the relationship between material removal and energy expenditure have been proposed. However, the assumptions leading to the specific energy appear to apply to slurry erosion. The specific energy allows calculation of the fluid mechanic variables and particle motions to be linked to the material removal rate. If the local power dissipated by the slurry particles at the wall can be calculated, the erosion depth will also be known from the specific energy of the slurry and construction materials.

In addition to the overall flow pattern in pumps, trailing vortices can produce severe local erosion (Tuzson 1999). When trailing vortices are stretched, the swirl velocity and centrifugal acceleration within the vortex can considerably intensify, and the vortex can grind out deep groves in the streamwise direction. Such trailing vortices are difficult to detect in clean water, but their presence becomes very evident from the erosion pattern of pumps from abrasive solid particles in water.

EROSION RATE ESTIMATES

The calculation of fluid flow velocities in a rotating impeller already presents considerable difficulty. The presence and motion of solid particles only increases the difficulty. Only computers offer the hope to achieve numerical results (Cader et al. 1993). The calculation of slurry particle velocities and concentrations in a pump impeller usually starts with calculation of the pure liquid flow. If low concentration can be assumed, and particle interactions and their effect on the liquid flow can be neglected, particle velocities can be calculated from the inertia and drag forces, which the fluid imposes on the particles. The concentration distribution is calculated from the particle flux due to the local particle velocities, based ultimately on the principle of conservation of mass. The concentration change in a fluid element results from the difference between the inflow and outflow of particles. However, special conditions must be imposed on the flow passage walls, where the particle concentration might reach its upper limit of maximum packing (Tuzson 1984). The assumption of low concentration loses its validity near the walls. Still, if the extent of the accumulation at the wall does not affect the liquid flow pattern significantly, valid results can be obtained. At high concentrations the analytical calculations become unmanageable. Mathematically, feasible calculation models require several speculative assumptions, which destroy the credibility of the calculation results.

Inertia and drag forces govern the motion of individual particles. A particle of diameter d in a fluid flow subjected to an acceleration of a experiences an inertia force F_I, and a drag force F_d given by the relations

$$F_I = \frac{\pi}{6} d^3 (\rho_s - \rho_l) a$$

$$F_d = C_d \rho_l \frac{\pi d^2}{4} \frac{(V - V_p)^2}{2}$$

The density of the particle and the liquid are ρ_s and ρ_l. The difference between the velocity of the fluid and the particle is $V - V_p$. The drag coefficient C_d depends on the Reynolds number formed from the velocity difference, the particle size, and the fluid kinematic viscosity v. In this case the Reynolds number takes on the form $Re = (V - V_p)d/v$. The coefficient of proportionality in the laminar range becomes $C_d = 24/Re$. For Reynolds numbers above 10^5, it is given by the expression $C_d = 0.2923[1 + (9.06/\sqrt{Re})]^2$ and remains about constant at $C_d = 0.15$ beyond.

Since the inertia and drag forces are vector quantities and have three components in the three directions of the coordinate system, which is being used, the acceleration and velocities must be taken in the corresponding directions. Drag forces are opposed to the direction of the velocity difference. Inertia forces act in the direction of the particle acceleration. Direct particle interactions at higher concentrations generally tend to increase the drag forces.

As a final step, the particle velocities, accelerations, and concentrations calculated near the passage walls must be linked to the material removal rate. In the case of large particles and low concentrations, impact erosion rate correlations can be used. Boundary layer turbulence is sometimes invoked to estimate impact angles and velocities (Chamkha 1994, Roco 1984). Turbulence calculations introduce further complications into the calculations and often become impractical. On the other hand, in the presence of small particles and high concentration, a sliding bed model is more likely and use of the specific energy concept can provide material removal rate estimates (Tuzson 1984). These complex calculations require a powerful computational fluid dynamic, a CFD type of computer program, modified specifically for suspended particles.

TYPICAL EROSION PATTERN IN SLURRY PUMPS

Some qualitative estimates can be made regarding the general erosion pattern in slurry pumps without detailed flow calculations. Erosion will predominate where the local acceleration is directed against the flow passage walls, where particles will accumulate, and where high velocities prevail. The direction of the acceleration will therefore be the guide to locate significant erosion. At the pump inlet, due to streamline curvature in the meridional, $(r-z)$ plane, the centrifugal acceleration will be directed toward the hub. As the flow accelerates in the circumferential direction, the centrifugal acceleration due to rotation comes into play. The resulting acceleration vector will gradually turn in the axial and then in the radial direction, but still oriented toward the hub. Slurry particles will tend to accumulate and slide along the hub.

On the blade-to-blade surface, accelerations in the radial and tangential directions are given by the following expressions, provided that the flow velocities change slowly in the circumferential direction. They correspond to the centrifugal and Coriolis accelerations in a rotating cylindrical coordinate system:

$$a_r = \frac{C_t^2}{r}$$

$$a_t = -C_r \frac{dC_t}{dr} + \frac{C_t}{r}$$

The direction of the tangential acceleration a_t is defined positive in the sense of rotation. Since the right-hand side of the corresponding expression is negative, the actual acceleration vector in the impeller points in the direction opposite to the sense of rotation. The angle of the resulting acceleration vector, β_a, is defined by

$$\tan \beta_a = \frac{a_t}{a_r} = -\frac{C_r(dC_t/dr + C_t/r)}{C_t^2/r}$$

In pump impellers the absolute tangential velocity C_t is positive and increases with the radius, and the radial velocity C_r is also positive. The angle β_a, measured here from the radial in the direction of rotation, is negative, as shown in Fig. 17.2. Therefore, the acceleration vector is always inclined in the direction opposite to the sense of rotation. According to the expression above, if the radial velocity C_r is small as in impellers with strongly backward-curved blades, the acceleration vector points almost in the radial direction, as shown in Fig. 17.2, and slurry particles would tend to accumulate on the suction side of the blades. Conversely, if the radial velocity is large, as in impellers with radial blades, for example, the acceleration vector is almost tangential, and the slurry particles would tend to accumulate on the pressure side of the blade.

Severe slurry erosion may also result from trailing vortices (Tuzson 1999). Tangential swirl velocities in vortices can spin up to very high values and generate high centrifugal accelerations perpendicular to the axis of the vortex filament. Such vortices can erode deep groves in the flow passage walls when solid particles are present in the fluid. Trailing vortices, being localized phenomena, are usually not identified in pump impeller flow calculations, and

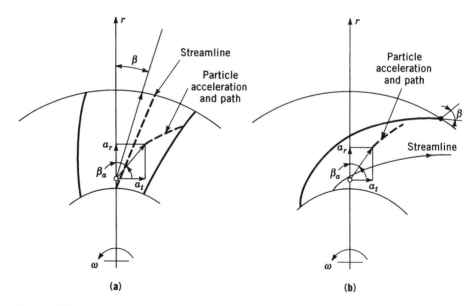

Figure 17.2 Particle acceleration and location of accumulation in a centrifugal pump impeller: (a) radial blades; (b) backward-curved blades.

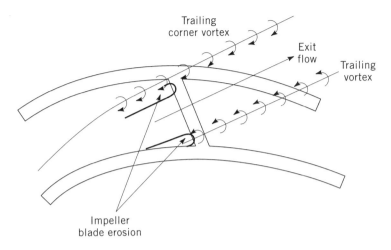

Figure 17.3 Impeller blade tip slurry erosion from trailing vortices.

with few exceptions, no simple calculation methods or analytical expressions have been developed. Trailing vortices are likely to appear at the blade leading edges at incidence, at the corners of the blades at the impeller exit on the hub and shroud side as shown in Fig. 17.3, and behind any protuberance at the outside faces of the impeller (Tuzson 1999).

Blade tip vortices occur at incidence, when the flow meets the leading edge at a slant, and the velocity, approaching the blade, has a component in direction of the leading edge. The separated region behind the tip of the blade forms a vortex, which trails away along the leading edge. In centrifugal pumps the leading edge usually slants toward the hub. The tip vortex detaches from the blade at the hub and is swept through the impeller along the hub. Near the point of detachment at the hub the vortex flow accelerates in its length direction, the vortex is stretched, and its swirl velocity increases. Intense erosion is often found near, or downstream of, the location where the blade leading edge meets the hub. The remedy for such an erosion consists of avoiding operation at off-design conditions. Large slurry pumps are often belt driven, which offers an opportunity to match the pump performance to the load characteristics by adjusting the rotational speed.

Trailing vortices at the corners of blades (Fig. 17.3), near the impeller exit, arise from an uneven angular momentum distribution across the trailing edge. Such vortices resemble wingtip vortices of aircraft, for example. A fundamental principle of fluid mechanics predicts that a vortex sheet will be spilled into the flow at the trailing edge of airfoils if the lift in the spanwise direction varies. Such a shear layer rolls up in wingtip vortices. In pumps, such trailing vortices result from nonuniform flow conditions at the blade trailing edges. Nonuniform conditions can be caused by the particular impeller geometry and blade design, or by boundary layers and flow separation along the hub and shroud surfaces. If the blades are heavily loaded near the impeller exit, and consequently the pressure difference between the pressure and suction sides is great, the trailing vortices near the impeller exit will accelerate and stretch, resulting in high rotational velocities and intense erosion. Therefore, if possible, the blades of slurry pumps should be unloaded near the exit, and flow separation in the impeller should be avoided. Uniform angular momentum distribution characterizes good pumps.

Erosion from trailing vortices can also appear on the outer hub and shroud faces of a slurry pump. The separation behind protruding bolts or indentations can generate trailing vortices. An approximate extent of the erosion can be obtained from a calculation model, which estimates the lift acting on the protuberance, since the lift is related directly to the vortex intensity and angular momentum of the trailing vortex (Tuzson 1999). The lift would then be estimated from the pressure forces acting on the protuberance. An exact numerical value of the erosion rate cannot be expected, but an order-of-magnitude estimate will become available. To minimize erosion, protuberances, cavities, or any geometry that could create a separated flow region should be avoided on the impeller of a slurry pump.

SLURRY EROSION TESTING

Erosion in slurry pumps can be minimized by selecting appropriate materials. The erosion resistance of materials must be determined by laboratory tests. The erosion rate in laboratory tests must be accelerated. Whereas in applications, significant erosion takes months or years, measurable erosion must be accomplished in the laboratory within hours. Two approaches are available to accelerate erosion rates: The erosion intensity must be increased significantly, or the material removed must be measured with extremely sensitive methods to minimize the need for deep erosion. An increase in erosion intensity can only be used with confidence if the relationship between material removal rate and intensity is well enough known over a sufficiently large range of values to allow a scaling of the results. Generally, it is not certain whether erosion rate increases exactly in direct proportion of the intensity of the forces or magnitude of the velocities over several orders of magnitude. A very accurate measurement of the material removed requires, that the original size and configuration of the sample be precisely known, and that the conditions of erosion be uniform and well defined.

Principal laboratory methods, specifically targeted at slurry erosion, are jet erosion testers, slurry pots and the Coriolis erosion tester, which are shown in Fig. 17.4 (Pagalthivarthi and Helmly 1992; Tuzson et al. 1984; Tuzson and Clark 1998). In jet erosion testing a high-velocity slurry jet strikes a flat specimen at some adjustable angle. In slurry pots a specimen rod of circular cross section, fixed to the end of an arm, is rotated around in a cylindrical container filled with slurry. In the Coriolis tester slurry is introduced into the center cavity of a rotor and exits through two radial channels (Tuzson et al. 1984). The specimen is

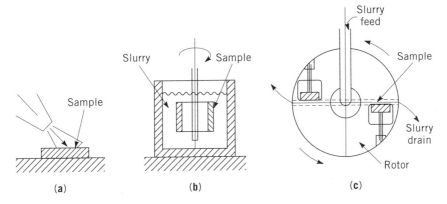

Figure 17.4 Laboratory methods targeted at slurry erosion: (*a*) jet erosion; (*b*) slurry pot; (*c*) Coriolis tester.

mounted into the wall of the radial channel facing the direction of rotation and is eroded by the slurry particles, which are pressed against the specimen face by the Coriolis acceleration.

In jet erosion tests and slurry pot tests the amount of material removed is determined by the weight loss. The samples are weighed before and after the test. In the case of the Coriolis erosion test the erosion groove is traversed using a profilometer, and the local erosion depth is determined to within a few microns. In jet erosion tests, the material that accumulates on the specimen surface tends to interfere with the incoming particles. The conditions of erosion vary on the specimen with the distance from the centerline of the jet. The weight loss of the specimen corresponds to an average erosion over the entire surface of the specimen. While maintaining the same test conditions, the erosion resistance of different materials can be compared, and can be ranked; a quantitative relationship between the conditions of erosion and the local material removal rate remain only approximate (Mueller et al. 1978).

In general, somewhat similar comments also apply to slurry pot tests. However, in a laboratory experiment (Wong and Clark 1993), local material removal was measured precisely with a custom-built measuring instrument, and assuming a model of slurry flow around the specimen, calculated results have been compared with test data. The correlation was based on the impact erosion model and has been found to give good agreement. However, results for particles smaller than about 100 μm tend to deviate from the test data. In practice, slurry pots are only used to rank materials based on weight loss.

In the Coriolis erosion tester (Tuzson 1984; Tuzson and Clark 1998), a relatively small batch of slurry, from an overhead tank, passes through the rotor in one pass. At rotor speeds of 6000 rpm a measurable groove is worn into the specimen within a few minutes. From the rotational speed and the radial velocity of the slurry flow, the local Coriolis force and sliding velocity can be estimated, which also defines the dissipated energy. The local erosion depth is determined from profilometer measurements to within a few microns. From these measurements the specific energy of the sample material can be calculated. The well-defined test conditions and accurate wear measurements allow closely reproducible results and good capability in distinguishing between the erosion resistance of a variety of materials (Tuzson and Clark 1998; Xie et al. 1999).

REFERENCES

Addie, G. R., Pegelthivarthi, K. V., Visintainer, R. J. (1996): *Centrifugal Slurry Pump Wear Technology and Field Experience*, ASME FED Vol. 236, pp. 703–715.

Beckmann G. (1980): A Theory of Abrasive Wear Based on Shear Effects in Metal Surfaces, *Wear*, Vol. 59, pp. 421–432.

Beke, B. (1964): *Principles of Comminution*, Akadémiai Kiado, Budapest.

Brach, R. M. (1984): Friction, Restitution and Energy Loss in Planar Collision, *ASME Journal of Applied Mechanics*, March, pp. 164–170.

Brach, R. M. (1988): Impact Dynamics with Application to Solid Particle Erosion, *International Journal of Impact Engineering*, Vol. 7, No. 1, pp. 37–53.

Brach, R. M., Editor (1991): Particle Surface Collisions: Erosion and Wear, in *Mechanical Impact Dynamics*, Wiley, New York.

Cader, T., Masbernat, O., Roco, M. C. (1993): Two-Phase Velocity Distributions and Overall Performance of a Centrifugal Slurry Pump, in *Pumping Machinery Symposium*, ASME FED Vol. 154, pp. 177–185. Also *ASME Journal of Fluids Engineering*, June, pp. 316–323.

Chamkha, A. J. (1994): Boundary Layer Theory for a Particulate Suspension, *ASME Journal of Fluids Engineering*, March, pp. 147–153.

Chen, T.-Y., Walawender, W. P., Fan, L. T. (1980): The Solid Flow Properties, *AIChE Journal*, Vol. 26, No, 1, pp. 31–36.

Clark, H. M., Wong, K. K. (1995): Impact Angle Particle Energy and Mass Loss in Erosion by Dilute Slurry, *Wear*, Vol. 186/187, pp. 454–464.

Hashish, M. (1984): A Modeling Study of Metal Cutting with Abrasive Water-Jets, *ASME Journal of Engineering Materials and Technology*, January, pp. 88–100.

Maan, N., Broese Van Groenou, A. (1977): Low Speed Scratch Experiments on Steels, *Wear*, Vol. 42, pp. 365–390.

Mueller, J. J., Wright, I. G., Davis, D. E. (1978): Erosion Evaluation of Materials for Service in Liquefaction Coal Slurry Feed Pumps, *Proceedings of the 3rd International Technical Conference on Slurry Transportation*, Las Vegas, Nev., March 29–31, pp. 107–116.

Mulhearn, T. O., Samuels, L. E. (1962): The Abrasion of Metals: A Model of the Process, *Wear*, Vol. 5, pp. 478–498.

Pagalthivarthi, K. V., Helmly, F. W. (1992): Applications of Material Wear Testing to Solids Transport via Centrifugal Slurry Pumps, in *Wear Testing of Advanced Materials*, ASTM Paper STP 1167, pp. 114–126.

Peterson, M. B., Winer, W. O. (1980): *Wear Control Handbook*, ASME, New York.

Roco, M. C., Nair, P., Addie, G. R., Dennis, J. (1984): Erosion of concentrated Slurries in Turbulent Flow, in *Liquid-Solid Flows and Erosion Wear in Industrial Equipment*, New Orleans, La., February 12–16, ASME FED Vol. 13, pp. 69–77.

Suzuki, M., Makino, K., Yamada, M., Iinoya, K. (1981): A Study on the Coordination Number in a System of Randomly Packed, Uniform-Sized Spherical Particles, *International Chemical Engineering*, July, pp. 482–488.

Truscott, G. F. (1972): *A Literature Survey on Abrasive Wear in Hydraulic Machinery*, BHRA, Cranefield, Tenn., October. Also *Wear*, Vol. 20, 1972, pp. 29–50.

Tuzson, J. (1984): Laboratory Slurry Erosion Tests and Pump Wear Rate Calculations, *ASME Journal of Fluids Engineering*, June, pp. 135–140.

Tuzson, J. (1999): Slurry Erosion from Trailing Vortices, *Fluids Engineering Division Summer Meeting*, San Francisco, July 18–23, FEDSM99-7794.

Tuzson, J., Clark, H. M. (1998): The Slurry Erosion Process in the Coriolis Erosion Tester, ASME Paper SM98-5144, *Fluids Engineering Division Summer Meeting*, Washington, D.C., June.

Tuzson, J., Lee, J., Scheibe-Powell, K. A. (1984): Slurry Erosion Tests with Centrifugal Erosion Tester, in *Liquid–Solid Flows and Erosion Wear in Industrial Equipment*, New Orleans, La., February 12–16, ASME FED Vol. 13, pp. 84–87.

Wellinger, K., Uetz, H. (1955): Gleitverschleiss, Spülverschleiss, Strahlverschleiss unter der Wirkung von körnigen Stoffen, *VDI-Forschungsheft*, Vol. 449, Pt. B, No. 21.

Wong, K. K., Clark, H. M. (1993): A Model of Particle Velocities and Trajectories in a Slurry Pot Erosion Tester, *Wear*, Vol. 160, pp. 95–104.

Xie, Y., Clark, H. M., Hawthorne, H. M. (1999): Modeling Slurry Particle Dynamics in the Coriolis Erosion Tester, *Wear*, Vol. 225/229, pp. 405–416.

Zenit, R., Joseph, G. G., Hunt, M. L. (1999): The Coefficient of Restitution for Liquid Immersed Collisions, *Proceedings of the 3rd ASME/JSME Joint Fluids Engineering Conference*, San Francisco, July, FEDSM99-7793.

18

EJECTOR PUMP SYSTEMS

SYSTEM DESCRIPTION

In many installations the suction head available from the application would be insufficient to satisfy the requirement of the pump. Such cases are encountered, for example, in well pumps where the water table is considerably below the surface. Vertical pumps with a long shaft or with a submersible electric motor can be used to lower the pump inlet below the water table level. Another, inexpensive simple solution consists of boosting the pump inlet pressure with an ejector, which entrains the inlet flow using a high-pressure side-stream from the pump. Ejectors are simple, inexpensive, rugged, and have no moving parts. The arrangement also has the advantage that the pump remains aboveground and is easily accessible (Elger et al. 1991; Hansen and Kinnavy 1965; Nece 1968; Sanger 1970; Stepanoff 1957).

A typical arrangement is shown in Fig. 18.1. The pump inlet head H_0 is shown positive for the purposes of calculation but can also be negative with respect to ambient. The pump produces a total head rise H_p. The ejector is lowered to a depth L below the surface and is submerged H_2 below the water level. The ejector pumping system delivers a net flow of Q_2 at a total head of $H_p + H_0$ to the user. The pump also supplies a flow rate of Q_1 to the primary jet of the ejector, which produces a total discharge head of H_d and a combined flow rate of $(Q_1 + Q_2)$ at the exit of the ejector.

Several cases can limit system operation. The static head at the mixing tube inlet H_{m1} might fall below the vapor pressure of the water, and the ejector may cavitate. When the output head of the ejector becomes insufficient to pressurize the pump inlet, the inlet head H_0 drops below the NPSHR of the pump, and the pump will cavitate.

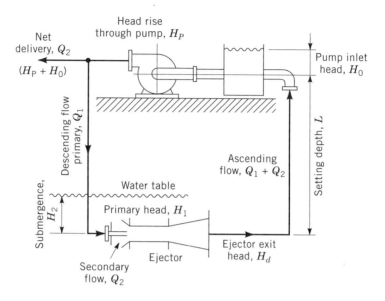

Figure 18.1 Ejector pump system.

SYSTEM CHARACTERISTICS

With the help of Fig. 18.1, expressions can be written for the ejector primary total inlet head H_1 and total delivery head H_d:

$$H_1 = H_p + H_0 + L - \text{loss1}$$
$$H_d = H_0 + l + \text{loss2}$$

Here loss1 and loss2 are the pipe friction losses in the descending and ascending pipes. As will be seen, the characteristics of the ejector can be represented by the dimensionless diagram relating a head ratio R_h and a flow rate ratio R_q with the ratio of cross-sectional areas of the primary nozzle and the mixing tube R as a parameter.

$$R_h = \frac{H_d - H_2}{H_1 - H_d}$$

$$R_q = \frac{Q_2}{Q_1}$$

In this expression, $H_d - H_2$ is the head rise of the secondary stream and $H_1 - H_d$ is the head loss of the primary stream. The product of the two quantities represents the fraction of the total energy transferred from the primary stream to the secondary stream, and therefore stands for the efficiency of the ejector:

$$\eta = R_h R_q$$

EJECTOR

A typical ejector, shown in Fig. 18.2, consists of a primary inlet nozzle exhausting axially into a cylindrical mixing section several diameters long. The secondary stream enters parallel through the annulus surrounding the primary nozzle. The high-speed primary and low-speed secondary streams mix in the mixing tube. A diffuser following the mixing tube slows the discharge stream and recovers some of its kinetic energy.

It can be assumed that the static pressure at the mixing tube inlet, where the primary and secondary streams merge, is the same for both streams. This assumption may not apply to unusual cases (Elger et al. 1991). It is conceivable that periodic vortices appear in the shear layer separating the two streams, which distort the pressure distribution. Still, assuming an identity of the pressures remains a good approximation. Indeed, the two streams are straight and parallel; therefore, no transverse pressure gradient can be present. The velocities of the two streams at that location will correspond to the head difference between the total or stagnation head and the static head at the mixing tube inlet and can be calculated from Bernoulli's equation for each stream. They can also be expressed by the corresponding flow rates divided by the cross-sectional areas. By the end of the mixing tube the two streams would have been fully mixed. The total head will be the sum of the static and velocity head at the mixing tube exit. A portion of the velocity head will be converted to static head in the diffuser following the exit of the mixing tube. The analytical problem of defining the ejector characteristics consists of finding an expression for the static head rise from the mixing tube inlet to its exit (Hill 1967).

Since all momentum and pressure forces in the mixing tube act in the same direction, the momentum equation, which is equivalent to the balance of forces in the axial direction, can be applied to a control volume defined by the inlet and exit cross sections and the cylindrical walls of the mixing tube. Wall friction forces will be neglected. Only the pressure forces on the inlet and exit cross sections and the momentum of the two streams arriving and of the mixed stream

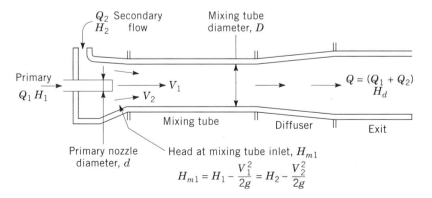

Figure 18.2 Ejector.

leaving need to be taken into account. Also using the continuity equation, the static head rise from the mixing tube inlet to the exit is finally given by

$$H_{m2} - H_{m1} = \frac{R - R^2}{g} \left(\frac{Q_1}{R} - \frac{Q_2}{1-R} \right)^2$$

In this expression R is the ratio of the primary nozzle cross-sectional area and the cross-sectional area of the mixing tube, the area ratio. The quantities Q_1/R and $Q_2/(1-R)$ correspond to the primary and secondary stream velocities. It will be noted that if the secondary flow rate becomes zero, the expression corresponds to the head recovery from a sudden expansion of the primary jet. It is also evident that the expression would remain the same if the secondary flow were to enter through the central nozzle and the primary jet through the periphery. The two streams are interchangeable, and only the difference in their velocities enters into the expression. For the expression to be valid, the mixing tube has to be long enough to assure that the exit velocity becomes uniform.

Having expressions for the static heads and all velocities at the mixing tube inlet and exit, the ejector performance characteristics can be expressed in terms of the total heads of the streams, in dimensionless form. In calculating the discharge total head, the assumption is made that only 65% of the mixing tube exit velocity head can be counted; 35% is lost in the diffuser. Such an assumption leads to good agreement with published results (Elger et al. 1991; Hansen and Kinnavy 1965). Writing the final form of the ejector performance characteristics in terms of the dimensionless ratios R_h and R_q, the expression becomes

$$R_h = \frac{H_d - H_2}{H_1 - H_d} = \frac{A_1 R_q^2 + B_1 R_q + C_1}{A_2 R_q^2 + B_2 R_q + C_2}$$

$$R_q = \frac{Q_2}{Q_1}$$

$$A_1 = \frac{1}{(1-R)^2} - k - \frac{2R}{1-R} \qquad A_2 = k + \frac{2R}{1-R}$$

$$B_1 = 4 - 2k \qquad B_2 = -4 + 2k$$

$$C_1 = -k - \frac{2(1-R)}{R} \qquad C_2 = k - \frac{1}{R^2} + \frac{2(1-R)}{R}$$

$$k = 0.65$$

This function is plotted in Fig. 18.3 for a range of variables that correspond to practical applications. The ejector efficiency is not shown separately but can easily be calculated from the product of the two dimensionless variables.

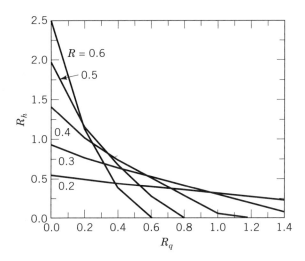

Figure 18.3 Ejector characteristics.

EJECTOR SPECIFICATIONS AND SELECTION

Application specifications will prescribe the net flow rate from the pump, Q_2, and the head $H_p + H_0$. The setting depth L and submergence H_2 will also be given.

The ejector geometry, in particular the nozzle and mixing tube diameters, and the total flow rate required from the pump, can be determined with the help of the dimensionless ejector performance characteristics. The head ratio will be a function of the application variables and can be obtained by substituting the expressions for the variables H_d, H_1, and H_2:

$$R_h = \frac{H_d - H_2}{H_1 - H_d} = \frac{H_0 + L - H_2 + \text{loss2}}{H_p - \text{loss1} - \text{loss2}}$$

If the pipe friction head losses loss1 and loss2 and the pump inlet head H_0 are relatively small compared with the setting depth L, the head ratio will roughly correspond to the ratio of the depth of the water table, $L - H_2$, and the pump head rise H_p:

$$R_h = \frac{L - H_2}{H_p}$$

For a first trial design the pump inlet head H_0 can be chosen zero and the pipe friction losses can be set at a few feet. With this value of the head ratio, a first choice of the area ratio R and the flow rate ratio R_q can be made from the ejector characteristics. Optimal conditions will be found around $R_h = 0.35$ to 1.0 and

$R_q = 1.0$ to 0.35, where ejector efficiencies will be in the 30s, as can be seen from their product.

Knowing the net flow required from the pump Q_2, the flow rate of the primary Q_1, and the total flow required from the pump, $Q = Q_1 + Q_2$ can be calculated from the flow ratio:

$$Q = Q_2\left(1 + \frac{1}{R_q}\right) \qquad Q_1 = \frac{Q_2}{R_q}$$

These values of the flow rates in the descending and ascending pipe branches can be used to calculate approximate values for the pipe losses loss1 and loss2, and corrected values for the area ratio and the flow rate ratio can be selected.

The exact diameter d of the primary nozzle of the ejector is determined by the primary flow rate Q_1 and the flow velocity, which in turn depends on the head difference $(H_1 - H_{m1})$ between the head supplying the nozzle and the static pressure head at the mixing tube inlet. Since the secondary flow rate Q_2 will depend in turn on the head difference $(H_2 - H_{m1})$ between the submergence and the static pressure head at the mixing tube inlet, the primary nozzle area can be expressed in terms of the head difference $(H_1 - H_2)$ by eliminating H_{m1} and introducing the area ratio R and the flow rate ratio R_q:

$$H_1 = H_p + H_0 + L - \text{loss1}$$

$$\frac{\pi}{4}d^2 = \frac{Q_1}{[2g(H_1 - H_2)/1 - R_q^2 R^2)]^{1/2}}$$

The mixing tube diameter D follows immediately from $D = d/R^{1/2}$. The mixing tube length is usually 8 to 10 times the mixing tube diameter. At this point the head at the mixing tube inlet H_{m1} can be checked to ascertain that no cavitation will take place:

$$H_{m1} = H_1 - \frac{Q_1^2(4/\pi d^2)^2}{2g}$$

With these values the major dimensions of the ejector have been defined.

EJECTOR SYSTEM PERFORMANCE MAP

Calculation of the entire performance characteristics of the ejector pump system may have to proceed by two or more steps of approximation since the experimental ejector characteristics and the pump head–flow curve can be too complex to be represented by analytical expressions, and therefore are often plotted in the form of curves. Typical pump and ejector performance curves, and their relationship in an ejector pump system, are shown in Fig. 18.4.

EJECTOR SYSTEM PERFORMANCE MAP 241

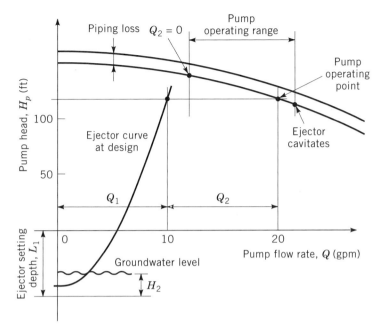

Figure 18.4 Ejector operating diagram.

The primary nozzle diameter d and the area ratio R of the ejector have been selected. In the approach presented here, the flow ratio R_q and the head ratio R_h will have been chosen from the ejector characteristics corresponding to the area ratio R, and the required pump head H_p and flow rate $Q = Q_1 + Q_2$ will be determined. The following three equations are used to eliminate H_0 and H_1 and to arrive at an equation in terms of the ejector ratios and the pump head–flow relationship:

$$H_1 = H_2 + \frac{Q_1^2(4/\pi d^2)^2}{2g}(1 - R_q^2 R^2)$$
$$H_0 = R_h(H_p - \text{loss1} - \text{loss2}) - L + H_2 - \text{loss2}$$
$$H_1 = H_p + H_0 + L - \text{loss1}$$

By rearranging, the expression is brought to the form

$$\frac{H_p - \text{loss1} - \text{loss2}}{Q^2} = \frac{(4/\pi d^2)^2}{2g} K_R$$

$$K_R = \frac{1 - R^2 R_q^2}{(1 + R_h)(1 + R_q)^2}$$

If the ejector head and flow ratios are selected from the performance curve of the ejector, the pump head divided by square of the flow rate can be calculated. The separate values of the pump head and flow rate will depend on the characteristics of the particular pump that is being used and can be determined from the head–flow curve of the pump.

If the losses can be assumed to be proportional to the square of the flow rates in the respective pipes, the expression above can be modified to take into account the effect of the flow rate variation on the losses:

$$\text{loss1} = k_1 Q_1^2 = \frac{K_1}{(1+R_q)^2} Q^2$$

$$\text{loss2} = K_2 Q^2$$

$$\frac{H_p}{Q^2} = \frac{H_p - \text{loss1} - \text{loss2}}{Q^2} - \frac{k_1}{1+R_q} - k_2$$

The primary Q_1 and secondary flow rate Q_2 can be calculated from the total flow rate Q and the flow rate ratio R_q:

$$Q_1 = \frac{Q}{1+R_q}$$

$$Q_2 = Q \frac{R_q}{1+R_q}$$

The practical operating range of the pump will have to remain between the extreme operating points of the ejector, $H_h = 0$ and $H_q = 0$. It may not be possible to reach these points because the pump inlet or the ejector inlet may cavitate. The corresponding negative pressure heads must stay above the vapor pressure of the water and are given by

$$H_0 = R_h(H_p - \text{loss1} - \text{loss2}) - L + H_2 - \text{loss2}$$

$$H_{m1} = H_1 - \frac{Q_1^2 (4/\pi d^2)^2}{2g}$$

The pump will operate over a limited portion of its total range. The pump flow rate Q will still be considerable when the secondary flow rate Q_2, which corresponds to the net flow delivered by the system, becomes zero. Similarly, the pump head H_p will remain high when the ejector ceases to pressurize the pump inlet.

DESIGN EXAMPLE

As an example, an ejector system will be calculated for the following input variables: $H_p + H_0 = 130$ ft (40 m), $H_0 = 0$, $Q_2 = 10$ gpm $= 38.5$ in³/s (0.000631 m³/s), $L = 60$ ft (18.3 m), $H_2 = 20$ ft (6.1 m), loss1 $= 5$ ft (1.5 m), and loss2 $= 5$ ft (1.5 m).

The following variables are calculated in sequence from the relationships given above. Since the head is given in feet and the flow rate in cubic inches per second, it is expedient to convert the head to inches in numerical calculations. Obviously, any measurement system can be used, SI or English, as long as consistent units are chosen. The expressions here presented do not contain constants, which are not dimensionless. The only exception might be the gravitational acceleration $g = 32.2$ ft/sec² (9.8 m/s²).

head ratio of ejector, $Rh = \dfrac{60 - 20 + 5}{130 - 5 - 5} = 0.375$

ejector area ratio, $R = 0.3$

flow rate ratio, $R_q = 1.0$

total pump flow rate, $Q = 77$ in³/sec (0.001262 m³/s)

primary flow rate, $Q_1 = 38.5$ in³/sec (0.000631 m³/s)

secondary flow rate, $Q_2 = 38.5$ in³/sec (0.000631 m³/s)

total pressure of the primary, $H_1 = 130 + 60 - 5 = 185$ ft (56.4 m)

cross section of the primary nozzle, $\dfrac{\pi d^2}{4} = 0.02969$ in² (0.191 cm²)

primary nozzle diameter, $d = 0.1944$ in. (0.494 cm)

mixing tube diameter, $D = 0.355$ in. (0.902 cm)

Note that the ejector efficiency is 37.5%, the highest for an area ratio of 0.3. With these measurements the system geometry is fully defined. One operating point has been determined on the head–flow curve of the pump, which should be close to the best efficiency point of the pump. Since the flow rates are known, the descending and ascending pipe sizes can be selected, and the head losses loss1 and loss2 can be verified. If they differ greatly from the assumed 5 ft (1.5 m), the design process can be repeated. Obviously, the losses should be kept as low as possible.

To calculate performance at other operating points, the head–flow characteristics of the pump would be needed. For the sake of illustration it will be assumed here that the pump characteristics is given by the correlation

$$H_p - \text{loss1} - \text{loss2} = 150 \text{ ft} - \dfrac{0.005 \text{ ft}}{\text{in.}^6/\text{sec}^2}(Q \text{ in}^3/\text{sec})^2$$

$$H_p - \text{loss1} - \text{loss2} = 46 \text{ m} - \dfrac{5.1 \times 10^6 \text{ m}}{\text{m}^6/\text{s}^2}(Q \text{ m}^3/\text{s})^2$$

This equation corresponds to an inverted parabola with its origin at $Q = 0.0$, where the shutoff pressure would be 150 ft (46 m). The operating point $H_p - \text{loss1} - \text{loss2}) = 120$ ft. and $Q = 77 \text{ in}^3/\text{sec}$ $(0.001262 \text{ m}^3/\text{s})$ lies on the curve. By substituting the design values of the head and flow rate, or using the expression containing the nozzle area and the ejector head and flow rate ratios, one can verify that

$$\frac{H_p - \text{loss1} - \text{loss2}}{Q^2} = \frac{(4/\pi d^2)^2}{2g} K_r = 0.02024 \text{ ft}/\text{in}^3/\text{sec})^2$$

$$K_r = \frac{1 - R^2 R_q^2}{(1 + R_h)(1 + R_q)^2}$$

To further verify performance, the pump inlet pressure $H_0 = 0$, and the ejector inlet static pressure $H_{m1} = 3.6$ ft (1.08 m) can be calculated.

Considering now the extreme case when there is no secondary flow $R_q = 0$, the ejector characteristics for the area ratio of $R = 0.3$ shows that the head ratio becomes $R_h = 0.8$. The following variables are then calculated using the assumed pump head–flow relationship:

$$\frac{H_p - \text{loss1} - \text{loss2}}{Q^2} = 0.068 \text{ ft}/(\text{in}^3/\text{sec})^2 [7.73 \times 10^7 \text{ m}/(\text{m}^3/\text{s})^2]$$

$$Q = Q_1 = 45.32 \text{ in}^3/\text{sec}(0.000743 \text{ m}^3/\text{s})$$
$$Q_2 = 0$$
$$H_p = 150 \text{ ft } (45.72 \text{ m})$$
$$H_0 = 67 \text{ ft } (20.4 \text{ m})$$
$$H_1 = 272 \text{ ft } (83 \text{ m})$$
$$H_{m1} = 20 \text{ ft } (6.1 \text{ m}) = H_2$$

Now taking an operating point at flow rates larger than the design flow rate, the ejector head and flow ratios of $R_h = 0.26$ and $R_q = 1.2$ will be used:

$$\frac{H_p - \text{loss1} - \text{loss2}}{Q^2} = 0.01747 \text{ ft}/(\text{in}^3/\text{sec})^2 \ [1.96 \times 10^7 \text{ m}/(\text{m}^3/\text{s})^2]$$

$$Q = 81.7 \text{ in}^3/\text{sec } (0.001339 \text{ m}^3/\text{s})$$
$$Q_1 = 37.14 \text{ in}^3/\text{sec } (0.000609 \text{ m}^3/\text{s})$$
$$Q_2 = 44.57 \text{ in}^3/\text{sec } (0.00073 \text{ m}^3/\text{s})$$
$$H_p = 126.6 \text{ ft } (38.58 \text{ m})$$
$$H_0 = -12.1 \text{ ft } (-3.7 \text{ m})$$
$$H_1 = 169.5 \text{ ft } (51.66 \text{ m})$$
$$H_{m1} = 0 \text{ ft}$$

The flow rate cannot be increased beyond this operating point because the ejector will cavitate. The pump inlet pressure H_0 remains acceptable provided that the pump has an NPSHR value of at least -12 ft (-3.7 m). It will be noted that the operating range of the pump is limited by the ejector form $H_p = 150$ ft (46 m) and $Q = 45.32$ in^3/sec $= 11.8$ gpm (0.000743 m^3/s), to $H_p = 126.6$ ft (38.58 m) and $Q = 81.7$ in^3/sec $= 21.2$ gpm (0.001339 m^3/s $= 1.33$ L/s). While the delivery head of the system changes little, the delivered flow rate Q_2 varies from zero to 44.57 in^3/sec $= 11.6$ gpm (0.00073 m^3/s).

REFERENCES

Elger, D. F., McLam, E. T., Taylor, S. J., (1991): A New Way to Represent Jet Pump Performance, *ASME Journal of Fluids Engineering*, September, pp. 439–444.

Hansen, A. G., Kinnavy, R. (1965): *The Design of Water-Jet Pumps*, ASME Paper 65-WA/FE 31 and 32.

Hill, P. G. (1967): Incompressible Jet Mixing in Converging–Diverging Axisymmetric Ducts, *ASME Journal of Basic Engineering*, March, pp. 210–220.

Nece, R. E. (1968): The Suction Manifold as a Water Jet Pump, *ASME Journal of Basic Engineering*, June, pp. 316–318.

Sanger, N. L. (1970): An Experimental Investigation of Several Low-Area-Ratio Water Jet Pumps, *ASME Journal of Basic Engineering*, March, pp. 11–20.

Stepanoff, A. J. (1957): *Centrifugal and Axial Flow Pumps*, Wiley, New York. Reprinted by Krieger Publishing, Malabar, Fla., 1993, pp. 402–424.

19

EVALUATION OF NEW FLUID MACHINERY CONCEPTS

DEVELOPMENT CRITERIA

In new product development, an early decision needs to be made as to whether a new concept, incorporating new technology, is likely to lead to profitable business (Tuzson 1993). Ideally, incremental funding of research and development projects should be in proportion to the probability of success. The magnitude of risk should govern the commitment of funds. A major factor in the decision is the required development effort and corresponding cost on one hand, and the risk or probability of success on the other. Unfortunately little technical and economic information is available at the beginning of a development process that would help in making such a decision.

The approach proposed here consists of considering separately various elements of the concept or invention, and identifying those that are already in use in other applications. New inventions often become feasible provided that the performance range of certain components can be stretched beyond the range of its present use. Elements that are not in commercial use or those whose performance range needs to be extended significantly will require extensive technological development before work can proceed to product development. Product development must not be undertaken until the technology has been developed. To build hardware to "try out" an idea is rarely cost-effective.

Technology development consist of the acquisition and accumulation of experimental data and analytical design tools. Experiments, not tests, are performed for the specific purpose of accumulating empirical data that will be used in the design process. Analytical studies are pursued to arrive at valid design procedures. These tools must be suitable to design lines of products with a variety of specifications. Theoretically, a patent should teach the invention and enable those skilled in the art to design, in a straightforward manner, a working and properly functioning device. However, a patent is but a first step, a small—almost negligible—part of the development process leading to profits. In practice, considerable work remains to be done before sufficient knowledge is accumulated to allow the confident design of a practical product. This accumulation of knowledge, which is usually company confidential and is not available at universities or in the public domain, becomes the intellectual investment and capital of the enterprise, which assures it a competitive commercial position for years in the future. The real value, sold by outstanding companies, does not consists of the bare hardware but of the quality assurance and technical support that these companies provide with the product.

Product development consists of the application of analytical design tools and empirical design data to the design and development of a particular product, designed to specifications by a certain deadline within a projected budget. Specifications, deadlines, and budgets are prescribed by business considerations, which are affected by marketing, manufacturing, and financial factors beyond any technological issues.

The classification of fluid machinery usually includes, quite broadly, machines imparting energy to, or extracting energy from, a fluid stream. The definition comprises positive-displacement and fluid dynamic pumps and compressors, which are often incorporated into power-generating systems such as engines or gas turbines or into refrigeration and various chemical process systems. Even ejectors would be included. However, in this instance, the designation will be restricted to fluid machines handling incompressible fluids, in particular positive-displacement or fluid dynamic pumps.

ENERGY TRANSFER

Fundamental analytical studies arrived at the conclusion that efficient energy transfer to or from a fluid must involve an unsteady, periodic, cyclic process (Dean 1959). The need for such a process is evident in positive-displacement pumps. However, pressure fluctuations are also necessary in centrifugal and axial flow pumps. The rotating blades necessarily produce a periodic passage of high- and low-pressure regions when seen by a stationary observer at the impeller exit. Only they can assure high efficiencies.

Fluid machinery does exist that transfers energy by steady viscous shear such as ejectors (Cunningham 1974; Hansen et al. 1965; Huang et al. 1985; Sanger 1970) and shear pumps (Craford and Rice 1974; Rice 1965; Truman et al. 1978). Shear pumps consist of closely spaced, parallel disks mounted on a rotating shaft which entrain liquid without blades, by the friction of the disk faces. However, the viscous friction required for energy transfer in these devices implies viscous energy losses. Practical ejector efficiencies rarely exceed 20%, and shear pump efficiencies remain considerably below the efficiencies of centrifugal pumps. Only unusual requirements, such as low noise or insensitivity to contamination, can justify their use. Ejectors have the advantage of no moving parts.

A hydraulic ram appears to operate on the principle of ejectors. In these devices slugs of a high-volume, low-head stream periodically compress batches of a lesser volume stream to higher pressure by the ram effect (Iversen 1975; Young 1997). The hydraulic ram cycles and uses the inertia of the driving fluid (not viscous forces) to transfer energy. It accelerates and then suddenly stops the flow in a long pipe, producing a short-duration, high-pressure pulse. Therefore, energy is transferred by periodic pressure fluctuations, not by viscous shear.

SPECIFIC SPEED

Specific speed N_s is a well-known figure of merit that characterizes the specifications of the fluid machinery application and can suggest the most appropriate type of machinery. It is dimensionless if consistent units, SI or English, are used; however, in industrial practice using English units of measurements, it is expressed in units of rpm, gpm, and feet.

$$N_s = \frac{NQ^{1/2}}{H^{3/4}}$$

In the English system, N is the shaft speed (rpm), Q the flow rate (gpm), and H the head (ft). Low specific speeds correspond to low flow rates and high pressures, while high specific speeds correspond to high flow rates and low pressures.

In turbomachinery selection and design, the specific speed of the application suggests the most efficient configuration: centrifugal, mixed flow, or axial turbomachines, depending on increasing specific speed. Best turbomachinery efficiencies correspond to a specific speed of about 2300 in English units (rpm, gpm, and ft). Efficiency declines rapidly with decreasing specific speed. Since the application sometimes specifies only the flow rate and head, leaving the shaft speed open, efficiency can be improved by choosing a shaft speed that changes the specific speed toward more optimal values.

Specific speed also governs the choice of hydraulic turbine configurations. At low specific speeds, pure impulse Pelton turbines give the highest efficiencies; centrifugal Francis turbines and axial Kaplan turbines are used at progressively higher specific speeds (Streeter 1961).

These trends result from fundamental physical principles, not from an arbitrary classification. At low specific speeds, high pressures, and low flow rates, leakage flows, increasing with pressure levels, become significant compared with the flow rate. Windage losses also become relatively important. At high specific speeds, high flow rates, and low pressures, the frictional flow losses predominate. The particular design configurations, optimal at the corresponding specific speed, minimize the prevailing losses.

Use of the specific speed to identify suitable fluid machinery configurations is not restricted to turbomachinery or fluid dynamic machinery (Balje 1962; Cartwright 1977). The lower shaft speeds favor positive-displacement machinery, as do high pressures and relatively low flow rates. The maximum head capability of turbomachinery is primarily a function of the tip speed, the circumferential velocity of the impeller, and is limited in commercial applications by the maximum speed of standard single-pole electric induction motors. Higher pressures, especially at small flow rates, call for screw, gear, or other rotary positive-displacement pumps, or ultimately for reciprocating piston pumps. Oil hydraulic pumps, handling pressures of up to 5000 psi and flow rates of just a few gpm, need to be of the positive-displacement type, preferably piston pumps (Tuzson 1978).

These analytical considerations are born out by the marketplace. Refrigeration and shop-air compressors best illustrate this trend. Approximately constant pressure ratings are set by industry standards, regardless of capacity. Flow rate is proportional to capacity. In turn the specific speed increases with flow rate and capacity. The smallest-capacity units sold commercially invariably use positive-displacement piston or rotary compressors, piston compressors being restricted to low speeds, on the order of 1000 rpm, and rotary compressors to electric motor speeds of 3600 rpm. At larger capacities, typically for commercial applications, screw compressors predominate, the speed of which is often geared up, since optimal efficiencies are reached at several thousand rpm. In very large industrial units, centrifugal or even axial compressors, driven at high speeds, are used. Small refrigeration systems have been proposed in the past using centrifugal compressors. However, these would require very high shaft speeds, on the order of 50,000 rpm, to reach sufficiently high specific speeds, which would assure satisfactory efficiencies. To date, no such systems are commercially competitive. Evidently, the specific speed of the application is a valuable guide to selection of the optimal fluid machinery configuration.

The most consistent technological trend through centuries consisted of increasing shaft speeds. An evident economic factor motivates the trend. If the shaft speed of a fluid-handling machine can be increased without significant modification to the hardware, the capacity of the device increases without additional cost and the price per unit capacity is reduced. Fluid machinery prices generally increase with the capacity, resulting in constant or slowly declining prices per unit capacity. Centuries ago, pumps were slow-running reciprocating units. Earlier hand drives gave way to waterwheel drives and were replaced by steam-driven linear reciprocating pumps in the early nineteenth century. Centrifugal pumps did not become practical until higher-speed drives appeared in particular electric motors.

FLUID MACHINERY TYPES

Fluid machinery types can be classed approximately in order of increasing specific speed. The highest pressures are encountered in oil hydraulic machinery and some chemical processes. Oil hydraulic applications typically require flow rates below about 100 gpm. Slow-running hydraulic motors use radial piston or cam-actuated piston devices, which can deliver very high torque (Olson 1966). Potentially high leakage and very high bearing loads dictate the use of high-viscosity lubricating oils. Operation with water would lead to immediate seizure. These devices are reminiscent of early radial aircraft engines, now obsolete. Similar engines have been proposed even recently (Decher 1984; Murray 1990) but they have not been developed to the point of commercialization.

Piston devices are the most popular, with a crank or connecting rod mechanism. However, free piston (Braun and Schweitzer 1973; Huber 1988), Scotch yoke (Stiller 1990; Weiss 1986), and swashplate devices (Robertson 1981; Scott 1981; Weber 1978; Yu and Lee 1986) have been proposed and tested, but are not commercially available. Axial piston-, swashplate-, or wobble plate-actuated devices did find mass-production applications in automotive air-conditioning compressors. They represent the preferred configuration for high-efficiency oil hydraulic pumps, often with variable capacity (Olson 1966; Tuzson 1978).

A great variety of rotary positive-displacement fluid machines can be found on the market (Nelik 1999). The German *Vereins Deutscher Ingenieure* (VDI) has published periodic reviews on positive-displacement pumps (VDI 1975). For low flow rates and moderately high pressures of a few hundred psi, gear pumps with fixed clearances are preferred (Scheel 1970). Minimum clearance widths are limited by manufacturing tolerance stackup, thermal expansion, and shaft runout. The gear tooth shapes can take several forms, as for example in the Roots-type blower. Internal and external gear arrangements can be distinguished. External gear types compete with piston pumps in oil hydraulic applications when fixed flow rates are acceptable. They can reach pressures of 1000 or 2000 psi with fixed clearances, and up to several thousand psi when pressure-balanced side plates are used to minimize leakage. Most inexpensive oil pumps, which form part of the integral lubrication system of engines, are of the internal gear type.

The Wankel family type of machines forms an exceptional configuration of internal gear arrangements (Yamamoto 1981). Leakage is minimized with sliding tip seals. Recently developed ceramic materials have overcome earlier apex seal wear problems. Wankel engines, mass produced by the millions by Mazda, represent the only engine type that has successfully challenged piston engines in automotive applications. The Wankel configuration has also been mass produced in a refrigeration compressor application.

The sliding vane type of fluid machinery configuration tends to overcome the tip leakage problem of fixed clearance devices (Karmel 1986; Peterson 1966). Lubrication of the sliding vanes remains a problem, however, and calls for the use of viscous oils. Such lubrication problems, especially at high temperatures, frustrate attempts at engine applications (Bassett 1990; Ewing 1982; Pennock and Beard 1988). A variable delivery configuration in which the outer cam can be displaced is on the market for oil hydraulic applications. Simple inexpensive positive-displacement pumps having rotors with integral vanes made of flexible rubber are on the market, but their limited life rules out demanding industrial applications.

Double- (Adkins and Larson 1970; Kaneko and Hirayama 1985) or single-screw (Chan et al. 1981; Zimmern and Patel 1972) pumps and compressors run with fixed clearances. With low-viscosity fluids, as in compressor applications, for example, high speeds on the order of 4000 to 8000 rpm are needed to minimize the effect of leakage. Pressure ratings remain a few hundred psi unless viscous fluids are handled or sealing oil is injected into those in compressor applications. The popularity of these pumps has increased because of recent advances in manufacturing tolerances and quality control. The elongated configuration of internal screw pumps presents an advantage in certain applications.

An unusual, very low specific speed pump is the pitot tube pump. It consists of a rotating water-filled case and a stationary radial pipe ending in a nozzle that points in the direction opposed to the sense of rotation on the periphery of the rotating case. The fluid, moving with the peripheral speed of the case, strikes the nozzle and can produce a maximum pressure equal to the stagnation pressure of the fluid, when no flow enters the nozzle.

Regenerative or side-channel turbomachinery occupies an intermediate position between positive-displacement and centrifugal machinery (Hollenberg and Potter 1979; Sasahara et al. 1980; Senoo 1956; Wilson et al. 1955). These pumps can deliver pressures in excess of those of centrifugal pumps and can remain competitive in selected applications, especially at low flow rates, but have generally low efficiencies.

The unusual approach of partial admission was attempted to extend centrifugal pump operation to low specific speeds (Wonsak 1963). In this device, intended for very low flow rates, the inlet flow is channeled to only one or two blade passages of a rotating radial impeller. At an appropriate circumferential location at the blade exit, a diffuser is positioned that receives the flow from the particular flow passage that happens to be filled with liquid. It is essentially a constant-pressure impeller in which the fluid is accelerated to high speed. The diffuser achieves a pressure increase by slowing the fluid. Efficiency will correspond to the diffuser pressure recovery, which might be in the range 50 to 70%, at most.

Centrifugal pumps reach acceptable efficiencies above a specific speed of about 500. Optimal efficiencies correspond to specific speeds of about 2300. The tip speed of centrifugal pumps, the product of impeller radius and the rotational speed, limits their pressure rating. Higher pressures demand multistage pumps. In smaller commercial units a single-pole synchronous motor speed of 3600 rpm is almost universal and results in a pressure rise per stage on the order of 25 to 200 psi, depending on impeller size. In process applications, variable-frequency drives are justified by better control. In larger commercial or industrial centrifugal pumps, lower speeds tend to be used. An extraordinary variety of special and custom-designed units have been developed in the past. At extremely high power densities and very high speeds and pressures, rotor-dynamic considerations begin to dominate feasibility (Cooper 1996).

Although radial inflow turbines are more efficient, inexpensive centrifugal pumps can also be used as power recovery turbines (Nelik and Cooper 1984). Power can be recovered in certain chemical processes and in water distribution systems by (instead of throttling) converting the energy from the reduction of the pressure of a fluid stream to electrical energy. Large hydroelectric water turbines are often designed to be reversible and act as pumps for pumped storage systems. In these cases the power produced by operating thermal power stations continuously during hours of low demand is stored by pumping water up into reservoirs, from where it is let down to produce peaking power during hours of high demand.

Mixed flow pump configurations, intermediate between centrifugal and axial flow configurations, overlap with centrifugal pumps at the lower end of their specific speed range. Mixed flow configurations are favored in the case of radial space limitations, such as in vertical multistage pumps which must fit in a certain well casing size. The factors affecting the choice of mixed flow pumps are similar to those for centrifugal pumps except for certain peculiarities of their characteristics: potential instabilities at reduced flow rates, for example.

Axial flow pumps become cost-effective at specific speeds of several thousand. They are usually single-stage designs since the need for higher pressures would lead to the choice of mixed flow pumps at better efficiencies rather than multistage axials. An axial stage or an inducer with spiral-shaped blades is sometimes used in front of centrifugal stages to minimize cavitation problems by producing a small pressure rise ahead of the impeller inlet. Axial pumps are most suitable for conveying very large amounts of fluids at very low pressures, such as flood control or dewatering applications.

FLUID MACHINERY COMPONENTS AND SYSTEMS

The feasibility and cost-effective functioning of fluid machinery depend on the entire system, not only on the fluid machine alone. Elements of the system assembly are often supplied by independent component manufacturers, all of whom are collectively held responsible by users for undisturbed operation of the system. The rating and reliability of the components or subassemblies must be matched to the fluid machine. Such components are, for example, bearings, seals, lubrication systems, and drives.

Seals

Seals rank as the most frequent service item in centrifugal pumps. Their purpose distinguishes two different kinds of seals. Seals preventing backflow from the impeller exit to the inlet are called *wear rings*. Since leakage detracts directly from the useful flow rate of a pump, minimizing leakage is indispensable. Wear rings, as their name indicates, also act as water-lubricated bearings and even allow metal-to-metal contact in case of emergency. Bearing forces from wear rings can significantly affect the rotor dynamics of the pump assembly and must be considered in rotor dynamic analysis.

Equally important are the *shaft seals*, which prevent leakage of the fluid being pumped from the pressure enclosure of the piping system and the intrusion of air into the piping system when subatmospheric pressures exist in the pipes. Careful design assures the balancing of the pressure forces on the mating stationary and rotating faces of the shaft seal, to minimize the clearance space between the seal faces. On the one hand, carbon or very hard materials are chosen to minimize friction and wear; on the other hand, flexibility is provided to allow the seal faces to accommodate misalignment, thermal distortion, and shaft runout. Proper alignment and installation of face seals are prerequisites of proper functioning and are important factors in seal failure.

The issue of leakage becomes particularly important in the case of environmentally harmful fluids. Absolute sealing can be assured when canned, hermetic electric motor and pump assemblies are used, in which the motor itself is immersed in the fluid being pumped and the fluid fills the air gap between the stator and the rotor. In centrifugal pumps that handle water containing suspended materials, special provisions need to be made to keep the suspended matter away from the shaft seal; otherwise, rapid deterioration and damage to the seal will occur. Seal contamination is prevented by continually flushing the seal with purified water. In some instances the contaminated water is purified by tapping into the pump outlet and piping a small stream to a hydraulic cyclone centrifugal separator, and from there to the vicinity of the seal.

In fixed-clearance positive-displacement pumps, manufacturing tolerance stackup, thermal and mechanical deformation, and shaft runout limit the minimum clearance width. A circular bore and piston that can be honed to very precise dimensions offers the most favorable geometry for sealing, and also acts as a bearing. Piston rings and O-rings have the most convenient geometry for best sealing. Self-adjusting pressure-balanced side plates in high-pressure oil hydraulic pumps actively minimize the side clearance width.

Bearings

The choice of the principal bearing types, hydrodynamic or rolling element, is dictated by the application. The bearing life required is the most important factor. Rolling element bearings have a well-defined life expectancy; hydrodynamic bearings do not. The life of rolling element bearings is adequate for aircraft or automotive applications, for example, which require approximately 2000 hours of operation, corresponding to about 100,000 miles of road travel for cars. The life requirement of consumer appliances typically remains below this level. Commercial machinery, including electric motors and directly coupled pumps, fall into a category requiring about 25,000 hours of life, about three years of continuous operation at 8760 hours per year. Rolling element bearings can still satisfy this requirement, provided that no side load is applied to the bearings, as it would be in belt, chain, or gear drives, and therefore the bearings only have to support the weight of the rotating parts. On the other hand, in power-generating machinery or process machines in practically constant operation, rolling element bearings will not last and sleeve bearings are mandatory. Power-generating machinery is designed for 40-year life with, at most, annual scheduled service. Process machinery applications often demand similar performance. Reliability becomes of primary importance because lifetime maintenance and service costs exceed the first cost in these applications.

Rolling element ball or roller bearings rely on rolling-contact Hertzian stresses to support the load. These periodic stresses lead eventually to fatigue failure of the ball or race. Slower-running and smaller rolling element bearings are lubricated with grease and are sealed. At very high speeds a fine mist lubrication is needed, since excessive oil or grease would increase the rolling friction and the heat generated. Consequently, a separate, active lubrication system is needed for such applications. Bearing life is proportional to the number of cycles or number of shaft revolutions, and typically depend on the third power of the bearing load. Evidently, an increase in bearing load reduces bearing life significantly. Successful applications with bearing lives in excess of 20,000 hours are very rare.

Hydrodynamic or sleeve bearings rely on the viscosity of the lubricating oil to produce a bearing force on the eccentrically positioned rotating shaft. These bearings must either be flooded with oil or need an external source of oil lubrication. The rotating shaft in a simple sleeve bearing must deflect to an eccentric position to produce a bearing force. However, the bearing force does not oppose the deflection directly but has a component in the transverse direction. These complex interactions between the direction of the deflection and the corresponding force lead to complex rotor dynamic effects and can even result in instabilities. Tilting pad hydrodynamic bearings have been conceived to counteract some of these problems. Multiple pivoting pads, arranged around the rotating shaft, self-adjust in response to the displacement and load applied.

In some machinery applications, externally pressurized bearings become possible, which, however, require a separate lubrication oil pump, typically of the gear type. The cost and complexity of an external lubricating system penalizes hydrostatic bearings. Such a separate lubrication system is more likely to be justified for larger machines. Hydrostatic pads can be made to support large loads and can be self-adjusting to assure and maintain close running clearances.

Bearing forces also arise from rotor unbalance forces, which depend on the unbalance or eccentricity tolerated in manufacturing, and increase with shaft speed. The life of an electric motor, mounted on ball bearings and balanced for normal duty at 3600 rpm, will rapidly deteriorate when operated at or above 5400 rpm. Since the load that a specific bearing has to support depends in a complex manner on the magnitude and distribution of the rotating masses, only a detailed rotor dynamic analysis can estimate the load bearing capability required from the bearing (Childs 1993).

During the last decade, electric bearings have been introduced, especially for larger industrial machinery (Habermann and Brunet 1985; Maslen et al. 1989). Bearing forces are produced, and the shaft is suspended without touching the bearing, by electromagnetic action of coils, arranged around the bearing periphery. Electric bearings have the advantage of precise control of the instantaneous bearing forces, and diagnosis of any anomalous behavior. Need for a separate lubrication system, and associated maintenance, is eliminated.

For very high rotor speeds of 50,000 to 100,000 rpm or higher, hydrodynamic bearings operating on air or some other gas, sometimes also called *foil bearings*, have been proposed and developed. Specialized applications, such as for dentist drills, exist, but their range of applicability and economic significance is very limited (Turner 1992).

Drives

The most frequent pump driver is an electric motor with a rotational speed of 3600, 1800, 1200, 900, or 720 rpm. The pump impeller is either mounted directly on the motor shaft or has a separate shaft and bearing assembly, in which case motor and pump are mounted on a frame and their shafts are connected with a coupling. A special procedure may be needed for startup, depending on the pump load. Very large electric motors, for example in power generation applications, may only be started twice in quick succession because of the danger of overheating.

When evaluating the cost and economic justification of an electric motor drive, the cost of needed accessories must also be counted. Electric motors need switchgear, starters, and circuit breakers, which sometimes cost as much as the motor itself. Larger motors may need special high-voltage transmission lines and voltage transformers or phase-correction equipment.

A variable-frequency variable- or high-speed electric motor drive for pumps may be desirable for three reasons. The maximum pressure produced by centrifugal pumps is limited primarily by the rotational speed. The need for complex, expensive multistage pumps could be avoided if a high-speed driver were available. The second reason is that better, more efficient control of the pump flow can be achieved with variable pump speed. In typical process applications a remotely controlled throttling valve follows the pump, which reduces the pump pressure by about 30% under normal operating conditions, to allow a suitable margin to increase or decrease the flow rate. Such throttling wastes 30% of the pumping power. Similar control can be achieved with variable speed without any losses. The third reason for the use of variable-speed motors is a better match of characteristics of the pump and the system load. Process equipment characteristics often change with time because of the deterioration of the equipment or because of modifications of the system. In such instances, the performance of existing equipment can sometime be adapted to the changed conditions by changing the rotational speed. For example, slurry pump wear can increase significantly if the pump is not operated at its rated condition. A change in speed can readjust pump performance to design conditions. Slurry pumps are often belt driven to provide an opportunity for speed adjustment. The cost of variable-frequency converters approaches $100 or more per kilowatt capacity, which is often prohibitive (EPRI 1985). The limited voltage capability of electronic components makes smaller-capacity, relatively lower voltage converters less expensive and more popular than high-powered units.

In remote or mobile applications such as irrigation, pumps are driven by internal combustion engines, most often diesels or spark-ignited natural gas engines. The shaft speed can be variable but is usually governed at a constant speed below about 2000 rpm. Diesel engines have relatively high shaft torque oscillations, which can damage couplings or cause bearing damage. Use of an internal combustion engine implies the necessity for several auxiliary systems: air filter, lubrication system, cooling water pump and heat exchanger, fuel pump, fuel tank, fuel controls, spark plugs or glow plugs, electric generator and associated electrical controls, batteries, starter motor, exhaust system, and safety equipment. The collective reliability, durability, cost, and proper functioning of all these components, regardless of their provenance, will determine the service that the fluid machine system provides to the user.

COST

The greatest uncertainty surrounds the cost of a novel machine. A detailed estimate of the manufacturing cost only becomes possible when detailed engineering drawings and a parts list are available, and when quotes have been obtained for all purchased items. Availability of this information implies that development has progressed to an advanced state and that considerable funds have already been spent on development. An early estimate of product cost is highly desirable.

An approximate first estimate of product cost can be obtained by calculating the anticipated weight per unit power of the machine, and by comparing this figure of merit with similar figures from existing competitive equipment. The underlying assumption is that the manufacturing cost of equipment with similar ratings is approximately proportional to the weight. This assumption is more likely to hold if an item is mass produced and therefore the cost of material accounts for a major portion of the manufacturing cost. However, new products can rarely be mass produced. Product introduction and market development take time. Sometimes years go by before sufficiently large sales volumes are reached. An early large capital investment in mass production tooling can result in high interest charges and in a negative cash flow for years, jeopardizing the timely recovery of the investment in development and startup. Sometimes it is easier to introduce a new kind of product based on new technology into a small but lucrative market segment with a custom-manufactured large capacity product, and switch to mass production after a positive cash flow has been achieved. A cost estimate based on weight per unit power has the advantage that commercial catalogs usually list the power rating and also the weight of the products. Therefore, it becomes relatively easy to establish the current status of commercially available products.

If drawings and parts lists are available, a detailed manufacturing cost estimate can be prepared. Unfortunately, a comparison with competitive products still remains difficult because only their sales price would be known, which includes confidential overhead charges, the amortization of development costs, sales cost, distributor markup, and profit. To get a valid comparison, the cost of the competitive product must be evaluated in exactly the same manner as the new product. Drawings and parts list must be prepared for the competitive product, also.

Ultimately, the price of a new product must be justified by the benefit it provides to the customer. Commercial applications usually require a payback period of about three years and corresponding amortization write-off. Obviously, product life must exceed the payback period. In these applications, purchase decisions are made by valuing low first cost more than lower operating cost or power consumption cost. Low cost and moderate efficiency are desired. Utility and process applications count on 20 or even 40 years of life and a correspondingly longer amortization period. Consequently, greater emphasis is given to low operating and service costs, which over many years can amount to a multiple of the purchase price.

A frequent mistake made when evaluating the feasibility of new products is comparing the present performance of commercial products with some future performance of the new product. Present commercial products will also be improved in future years. Often, manufacturers have already improved product prototypes on the shelf but must delay their introduction until the investment costs of present products are recovered. The obstacle to early introduction is not lagging technical development but an economic consideration. The staying power of existing products must not be underestimated.

REFERENCES

Adkins, R. W., Larson, C. S. (1970): *Basic Geometric Method in Helical Lobe Compressor Design*, ASME Paper 70-WA/FE-23.

Balje, O. E. (1962): A Study on Design Criteria and Matching of Turbomachinery, *ASME Journal of Engineering for Power*, January, pp. 83–102, 103–114.

Bassett, H. E. (1990): Rotary Compressor for Heavy Duty Gas Service, U.S. patent 4,960,371, October 2.

Braun, A. T., Schweitzer, P. H. (1973): *The Braun Linear Engine*, SAE Paper 730185.

Cartwright, W. G. (1977): Specific Speed as a Measure of Design Point Efficiency and Optimum Geometry for a Class of Compressible Flow Turbomachines, in *Scaling for Performance Prediction in Rotodynamic Machines*, Institution of Mechanical Engineers, New York, pp. 139–145.

Chan, Y., et al. (1981): The Hall Screw Compressor for Refrigeration and Heat Pump Duty, *International Journal of Refrigeration*, September, p. 275.

Childs, D. (1993): *Turbomachinery Rotordynamics Phenomena: Modeling and Analysis*, Wiley, New York.

Cooper, P. (1996): Perspective: The New Face of R&D, *ASME Journal of Fluids Engineering*, December, pp. 654–664.

Craford, M. E., Rice, W. (1974): Calculated Design Data for the Multiple-Disk Pumps Using Incompressible Fluid, *ASME Journal of Engineering for Power*, July, pp. 274–282.

Cunningham, R. G. (1974): Gas Compression with the Liquid Jet Pump, *ASME Journal of Fluids Engineering*, September, pp. 203–215, 216–226. Also March 1975, p. 133.

Dean, R. C., Jr. (1959): On the Necessity of Unsteady Flow in Fluid Machines, *ASME Journal of Basic Engineering*, March, pp. 24–28.

Decher, R. (1984): *The Britalus Brayton Cycle Engine*, ASME Paper 84-GT-258.

EPRI (1985): *Adjustable Speed Drives: Directory—Manufacturers and Applications*, Electric Power Research Institute, Palo Alto, Calif.

Ewing, P. (1982): *The Orbital Engine*, SAE Paper 820348.

Habermann, H., Brunet, M. (1985): *The Active Magnetic Bearing Enables Opimum Control of Machine Vibrations*, ASME Paper 85-GT-221.

Hansen, A. G., Kinnavy, R., Jett, A. V. (1965): *The Design of Water Jet Pumps*, ASME Paper 65-WA/FE-31 and 32.

Hollenberg, J. W., Potter, J. H. (1979): An Investigation of Regenerative Blowers and Pumps, *ASME Journal of Engineering for Industry*, May, pp. 147–152.

Huang, B. J., et al. (1985): Ejector Performance Characteristics and Analysis of Jet Refrigeration System, *ASME Journal of Engineering for Gas Turbines and Power*, July, pp. 792–802.

Huber, R. (1988): Free Piston Engines (in French), *Entropie*, Vol. 24, No: 141, May, pp. 3–24.

Iversen, H. W. (1975): An Analysis of the Hydraulic Ram, *ASME Journal of Fluids Engineering*, June, pp. 191–196.

Kaneko, T., Hirayama, N. (1985): Study of Fundamental Performance of Helical Screw Expander, *Bulletin of JSME*, September, pp. 1970–1977.

Karmel, A. M. (1986): A Study of the Internal Forces in a Variable-Displacement Vane Pump, *ASME Journal of Fluids Engineering*, June, p. 227.

Maslen, E., Hermann, P., Scott, M., Humphris, R. R. (1989): Practical Limits to the Performance of Magnetic Bearings, *Journal of Tribology*, April, pp. 331–336.

Murray, J. (1990): Rotocam Engine, *New Scientist*, August 4, p. 33.

Nelik, L. (1999): *Centrifugal and Rotary Pumps*, CRC Press, Boca Raton, Fla., 1999.

Nelik, L., Cooper, P. (1984): *Performance of Multi-Stage Radial-Inflow Hydraulic Power Recovery Turbines*, ASME 84 Paper WA/FM-4.

Olson, J. (1966): What IS New in Piston Pumps, *Proceedings of the 22nd National Conference on Fluid Power*, October, pp. 133–141.

Pennock, G. R., Beard, J. E. (1988): *A Variable Stroke Engine*, SAE Paper 880577.

Peterson, R. E. (1966): What IS New in Vane Pumps, *Proceedings of the 22nd National Conference on Fluid Power*, October, pp. 127–132.

Rice, W. (1965): An Analytical and Experimental Investigations of Multiple-Disk Turbines, *ASME Journal of Engineering for Power*, January, pp. 29–36.

Robertson, D. (1981): Revolver Barrel Engine: The New K-Cycle, *Science & Mechanics*, Winter, pp. 58–61, 101.

Sanger, N. L. (1970): An Experimental Investigation of Several Low-Area-Ratio Water Jet Pumps, *ASME Journal of Basic Engineering*, March, pp. 11–20.

Sasahara, T., Yamazaki, S., Tomita, Y. (1980): Researches on the Performance of the Regenerative Type Fluid Machinery, *Bulletin of the JSME*, December, pp. 2047–2054.

Scheel, L. F. (1970): A Technology for Rotary Compressors, *ASME Journal of Engineering for Power*, July, pp. 207–216.

Scott, D. (1981): Inclined Axes Design Simplifies Piston Engines/Compressors, *Automotive Engineering*, April, pp. 76–82.

Senoo, Y. (1956): A Comparison of Regenerative Pump Theories, *Transactions of ASME*, July, pp. 1091–1102.

Stiller (1990): The Stiller-Smith I. C. Engine, *3rd International New Energy Technology Symposium*, Planetary Association for Clean Energy, Hull, Quebec, Canada.

Streeter, V. L. (1961): *Handbook of Fluid Mechanics*, McGraw-Hill, New York.

Truman, C. R., Rice, W., Jankowski, D. F. (1978): Laminar Through-Flow of Varying Quality Steam Between Corotating Disks, *ASME Journal of Fluids Engineering*, June, pp. 194–200.

Turner, A. B. (1992): *Externally Pressurized Air-Bearing System with Applicability to Small Turbomachinery*, ASME Paper 92-GT-382.

Tuzson, J. (1978): Literature Sources on Positive Displacement Pump Design, *Proceedings of the 34th National Conference on Fluid Power*, November, pp. 399–402.

Tuzson, J. (1993): Evaluation of Novel Fluid Machinery Concepts, *ASME Fluids Engineering Summer Conference*, Washington, D.C., FED Vol. 154, pp. 383–386.

VDI (1975): *Verdrängungspumpen*, VDI, Berlin, pp. 542–546. Also (1973): pp. 493–496; (1972): pp. 399–403; (1969): pp. 838–840.

Weber, A. (1978): *Hydrostatische Axiallager in Axialkolbenmaschinen der Schrägachsenbauart*, VDI, Berlin, August, pp. 736–742.

Weiss, B. (1986): Advanced Technology Engine Receives Extension, *Metalworking News*, August, p. 150.

Wilson, W. A., et al. (1955): A Theory of the Fluid-Dynamic Mechanism of Regenerative Pumps, *Transactions of ASME*, November, pp. 1303–1316.

Wonsak, G. (1963): Die Strömung in einer partiell beaufschlagten radialen Gleichdruckkreiselpumpe, *Konstruktion*, Vol. 15, No. 3, March, pp. 99–106.

Yamamoto, K. (1981): *Rotary Engine*, Sankaido Co. Ltd., Tokyo.

Young, B. (1997): Design of Homologous Ram Pumps, *ASME Journal of Fluids Engineering*, June, pp. 360–365.

Yu, Z., Lee, T. W. (1986): Kinematic, Structural and Functional Analysis of Wobble-Plate Engines, *Journal of Mechanical Transmissions in Automotive Design*, June, pp. 226–236.

Zimmern, B., Patel, G. C. (1972): Design and Operating Characteristics of the Zimmern Single-Screw Compressor, *Purdue International Compressor Conference*.

20
COMPUTER PROGRAMS FOR PUMPS

COMPUTER CALCULATIONS

The computer programs discussed here are intended for use in solving flow equations for pumps. Computer programs predicting pump performance are discussed in the appropriate chapter. Computer programs for pump selection from commercial catalog databases, for pipe friction loss calculations, for water hammer calculation, or for empirical pump design formulas are not discussed here.

The actual fluid flow in pumps and related components is extremely complex. Calculations are performed by assuming a model, which is a simplified, imaginary configuration of the flow field. If the model is too simple, the calculation results will be trivial. If the model is too detailed, the calculation becomes too complicated. Ideally, a good model should be simple but still provide valid, useful calculation results. A complex and refined model is sometimes expected to give more precise results, which unfortunately, is not always the case. The complexity of the model should be matched to the need or economic justification for the effort required to obtain a solution.

Computer programs available today only verify the flow pattern in an existing pump geometry. The pump geometry is an input to the computer program and must be available. The pump flow passages must be designed before an impeller, diffuser, or volute flow calculation computer program can be executed. Only simplified, approximate calculation procedures are presented here, which give adequate, practical results. They are really hand calculations but can be done using faster computers. Pumps have to operate satisfactorily over a range of conditions that are not defined precisely or can change with time. In the face of such inherent uncertainties, the accuracy of calculated results does not need to be pushed to unreasonable limits. In some instances, or for more precise design calculations, these simple methods do not suffice: for example, when the flow becomes strongly three-dimensional. The problem of predicting the inlet flow pattern of centrifugal pumps presents such a case. Because of the curvature of the streamlines, pressure and velocity changes in all directions. A solution requires that the four basic partial differential equations be solved. Numerical methods must then be used, which are easiest to solve on high-speed computers.

Before the advent of high-speed computers, flow equations were generally based on the ideal potential flow model, which were solved either analytically, sometimes using conformal transformations, or by stepwise integration, which had to be performed using hand calculations or mechanical adding machines. Because of earlier limitations of computer capabilities, an approximate quasi-three-dimensional calculation method was later used. It consists of calculating flow in a pump impeller separately in the meridional hub-to-shroud and circumferential blade-to-blade directions, as shown in Fig. 20.1. The two calculations are linked by assuming averaged blade-to-blade conditions in the meridional calculations. Theoretically, calculations could be repeated with progressive corrections. Modern computers can execute a direct solution of the three-dimensional, partial differential equations, the Navier–Stokes equations, provided some means are included to handle turbulent fluctuations.

Two principal approaches are in use for executing complex flow calculations: the finite difference and finite element methods. In finite difference calculations, derivatives of the partial differential equations, which are based on infinitely small increments of the variables, are replaced by finite increments—hence the name *finite differences*. In finite element calculations, the forces acting on small volume elements of the flow are balanced directly. In either case, the flow region of interest is covered with a network of mesh points at which pressure and velocity

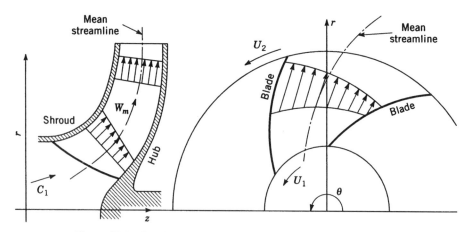

Figure 20.1 Quasi-three-dimensional velocity calculation in impeller.

are calculated by successive iterations. Unfortunately, because of the implied approximations, the calculations often do not converge, and despite many trial iterations, the desired solution is not approximated. Special calculation techniques can overcome these problems of convergence. Usually, a fine mesh with many mesh points, and lengthy iterations, are required. The calculation accuracy increases in proportion to the number of mesh points; the calculation time increases roughly with the square or cube of the number of mesh points. The assumptions regarding the inlet velocity distribution and the exit conditions, which constitute the boundary conditions of the mathematical model, also give trouble sometimes, in particular when inlet or exit flow recirculation is present.

Powerful computational fluid mechanic (CFM) computer programs are commercially available (Douglass and Ramshaw 1994; Lakshminarayana 1991; McNally and Sokol 1985; Muggli et al. 1997; Schachenmann and Muggli 1993). As input they require a precise numerical description of the flow boundaries, which determines the number of mesh points required. Sharp curvatures of the boundaries require a particularly fine mesh. It then becomes an economic consideration whether the time and expense of preparing and running the computer calculation is warranted by the value of the pump to be designed. Impeller calculations have been executed with several hundred thousand mesh points running on a workstation overnight. Such calculations are warranted for large, reversible pump-turbines, for example, which can have a performance penalty of $1 million per percentage point deficiency.

A key issue in the calculation of real flow pattern remains the handling of turbulence. Theoretically, an exact solution of the fundamental partial differential equations, the Navier–Stokes equations, will describe the instantaneous flow pattern perfectly. However, turbulent flow is unsteady, ever changing. Design requires that averaged, approximately steady values be calculated. The details of instantaneous fluctuations are of little interest. Consequently, some average representation of the turbulent fluctuations is needed, which although artificial, will be suitable for design purposes. Several such turbulence models have been proposed and are in use which can give satisfactory results, provided that they have been calibrated on test cases of similar flow patterns (Bradshaw et al. 1996; He and Walker 1995; Menter 1996).

REFERENCES

Bradshaw, P., Launder, B. E., Lumley, J. L. (1996): Collaborative Testing of Turbulence Models, *ASME Journal of Fluids Engineering*, Vol. 118, June, pp. 243–247.

Douglass, R. W., Ramshaw, J. D. (1994): Perspective: Future Research Directions in Computational Fluid Dynamics, *ASME Journal of Fluids Engineering*, Vol 116, June, pp. 212–215.

He, J., Walker, J. D. A. (1995): A Note on the Baldwin-Lomax Turbulence Model, *ASME Journal of Fluids Engineering*, Vol. 117, September, pp. 528–531.

Lakshminarayana, B. (1991): An Assessment of Computational Fluid Dynamic Techniques in the Analysis and Design of Turbomachinery, *ASME Journal of Fluids Engineering*, September, pp. 315–352.

McNally, W. D., Sokol, P. M. (1985): Review: Computational Methods for Internal Flows with Emphasis on Turbomachinery, *ASME Journal of Fluids Engineering*, Vol. 107, March, pp. 6–22.

Menter, F. R. (1996): A Comparison of Some Recent Eddy-Viscosity Turbulence Models, *ASME Journal of Fluids Engineering*, Vol. 118, September, pp. 514–519.

Muggli, F. A., Eisele, K., Casey, M. V., Gulich, J., Schachenmann, A. (1997): Flow Analysis in a Pump Diffuser, Part 2: Validation and Limitations of CFD for Diffuser Flows, *ASME Journal of Fluids Engineering*, December, pp. 978–985.

Schachenmann, A., Muggli, F. (1993): Comparison of Three Navier-Stokes Codes with LDA Measurements on an Industrial Radial Pump Impeller, in *Pumping Machinery Symposium*, ASME FED Vol. 154, pp. 247–252.

APPENDIX

This appendix contains FORTRAN computer listings for the programs LOSS3, INLET, IMPEL, MIXED, CUTS, BLADE, and EXREC. They are written and can be edited in DOS. A math processor and a FORTRAN compiler are needed to execute the programs, unless they are modified. The programs require that an input file, FILE.DAT, be set up containing input variables in free ASCII form. The programs generate several output files. The principal file, FILE.OUT, presents the calculated numerical data identified by their symbols. The output FILE.GRF lists blade coordinates as well as a reference circle corresponding to the impeller circumference, which are in free-form ASCII and are suitable for transfer to a plotting program. They can be edited before plotting. The individual calculation steps of the listings can be incorporated into and customized for more advanced computer systems. The calculation procedures contained in the computer programs are discussed in Chapters 11 and 12.

Calculations are done in English units. Pump geometry is in inches, velocities in ft/sec, and flow rates in gpm. (1 in. = 2.54 cm, 1 ft = 12 in. = 30.48 cm = 0.3048 m, 1 gall = 231 in^3 = 3785 cm^3 = 3.785 L = 0.003785 m^3.)

LOSS3 offers a very simple performance estimate for centrifugal and mixed flow pumps. It is described in detail in Chapter 10. The program provides quick, approximate performance estimates for alternative design trials. The calculation will not give correct results unless the loss coefficients are first calibrated on a similar pump. Reasonable adjustments in the input variables must be made by the designer to account for details such as nonuniform inlet flow or boundary layer blockage, for example. The output file LOSS.OUT prints the estimated head, efficiency, various head losses, and velocities for six selected flow rates.

INLET uses a very simple calculation to estimate the meridional velocity along the leading edge of impeller blades. Coordinates of the hub and shroud, the approximate radius of curvature at hub and shroud, and the leading-edge coordinates on hub and shroud need to be listed in the input file INLET.DAT, in addition to the flow rate.

IMPEL calculates blade coordinates on the hub and shroud with the same angular momentum distribution on all streamlines.

MIXED calculates impeller blade coordinates on hub, shroud, and mean streamlines for a blade shape that has inclined parallel straight-line generators conncting corresponding coordinate points on hub and shroud. Consequently, these blade shapes can be pulled individually from an impeller mold in the direction of the generators. The mean streamline coordinates, the angular momentum at the exit and the flow rate, must be listed in the input file MIXED.DAT. The calculated coordinates, the angular momentum distribution and the local blade angles, are listed in the output file MIXED.OUT. A plot of the coordinates transferred from the output file MIXED.GRF can verify whether the blades will pull.

CUTS prepares intersections of the blade shapes with equally spaced planes perpendicular to the axial direction. Such intersections are customarily shown on engineering drawings of impellers. The input file CUTS.DAT can be transferred from the MIXED blade coordinate calculation program.

BLADE calculates radii and centers of arcs of circles representing a simple two-dimensional blade in an axial projection. Inlet and exit blade angles and the radii at which the arcs are to join are specified in the input file BLADE.DAT.

MIXED is also used in an example to calculate the blade coordinates of the return channel or bowl of a multistage vertical pump. The calculation proceeds from the exit to the inlet, in the direction opposite to the direction of flow. The mean streamline coordinates, the flow rate and velocities at the return passage inlet, need to be input.

EXREC calculates the meridional velocity distribution along the blade trailing edge of mixed flow pumps at the onset of exit recirculation, and the corresponding flow rate. It assumes that at rated conditions uniform angular momentum is produced at the impeller exit.

```
NOMENCLATURE FOR LOSS3

INPUT FILE: LOSS.DAT
Free form input.
First line: Up to 40 characters for TITLE including
date.
Second line:QD, RPM,D1,DS,D2,B,B3,D4,AREA
 Third line:BEB1D,BEB2D,BEB3D,XNB,XNV,CDF,CSF,CRE,CVD
  (If XNV=1 a single stage volute pump is calculated)
  (If set to 0.0,CDF=00.005,CSF=0.005,CRE=0.001,CVD=0.5)
 Fourth line:Six flow rates to be calculated Q(I),I=6)
```

APPENDIX **269**

```
Q(I),gpm   Six values of the flow rate
QD,gpm     Nominal flow rate. Start of recirculation.
RPM        Rotational speed
D1,inch    Inlet shroud diameter
DS,inch    Shaft diameter
D2,inch    Impeller mean exit diameter
D3,inch    Diffuser or volute throat mean inlet diameter
D4,inch    Exit flange inside diameter
B,inch     Impeller exit width
B3,inch    Diffuser inlet width
BEB1D      Impeller inlet blade angle from meridional,
              degrees
BEB2D      Impeller exit blade angle from meridional,
              degrees
BEB3D      Diffuser inlet vane angle from meridional,
              degrees
XNB        Number of impeller blades
XNV        Number of diffuser vanes
CDF        Disk friction coefficient
CSF        Skin friction coefficient
CRE        Coefficient of diffuser or volute inlet loss
CVD        Coefficient of diffuser loss
AREA       Volume throat area
```

OUTPUT FILE: LOSS.OUT

First prints TITLE
Prints input: Head,ft., efficiency, theoretical hydraulic head,ft., losses,ft., velocities,ft/sec., leakage,gpm. for the six selected flow rates, gpm.

```
Q(I)       Flow rate,
DH         Head
EFF        Efficiency
DHTH       Theoretical hydraulic head
QDIN12     Impeller inlet incidence loss
QDSF12     Impeller skin friction loss
DQDIF      Impeller diffusion loss, if W1/W2.GT. 1.4
DQIN23     Diffuser inlet incidence loss
DQSF23     Volute skin friction loss
DQSF34     Diffuser skin friction loss
DQVD       Diffuser loss
DFH        Disk friction head loss, includes mechanical
              losses
DRCH       Recirculation head loss
```

270 APPENDIX

QL   Leakage loss

U1   Inlet rotational velocity
U2   Impeller tip speed
C1   Inlet absolute flow velocity
W1   Inlet relative flow velocity
WM1   Inlet relative, meridional velocity at correct incidence
W2   Impeller exit relative velocity
WM2   Impeller exit relative meridional velocity
CT2   Impeller exit absolute tangential velocity
CT3   Diffuser inlet absolute tangential velocity
C3   Diffuser inlet velocity
CQ3   Diffuser or volute throat velocity from Q/AREA
C4   Exit flange velocity

PLOTTING FILE: LOSS.GRF

Writes headings and six lines of Q(I),DH,DHTH,EFF and WORK=DF/EFF

```
C Program LOSS3.FOR - John Tuzson 7/22/97
C John Tuzson & Associates, 1220 Maple Ave. Evanston,IL
  60202

C Calculates head losses in pumps and head vs flow for
  six flow
C rates Q(I). Input file is LOSS.Dat: First line can have
  40
C characters for TITLE. Results are printed in
  LOSS.OUT.
C Data for plotting is in LOSS.GRF: Q(I),
  DH,DTHT,WORK,EFF.
C WORK=Head/Efficiency is work input for comparison with
  test data.
      DIMENSION Q(6)
      CHARACTER*40 TITLE
      OPEN(UNIT=50,FILE='LOSS.DAT',STATUS='OLD')
      OPEN(UNIT=60,FILE='LOSS.OUT',STATUS='UNKNOWN')
      OPEN(UNIT=70,FILE='LOSS.GRF',STATUS='UNKNOWN')
C QD=Design flow,DN rpm,D1 inlet dia,DS shaft dia,D2
  imp.exit dia,
C D3 diff.in dia, B imp.width,B3 diff.width, BEB1D
  inlet degree,
C from radial, BEB2D imp.out,BEB3D diff.in,XNB
  Nr.blades,XNV Nr.vanes.
```

```
C Volute = XNV=1, AREA volute throat,D4 exit flange dia,
  CDF disk
C friction coeff., CSF skin friction coeff., CRE
  recirculation coeff.
C CVD diffuser coeff., CIN diffuser/volute approach
  coeff.
      READ(50,*)TITLE
      READ(50,*)QD,DN,D1,DS,D2,D3,B,B3,D4,AREA
      READ(50,*)BEB1D,BEB2D,BEB3D,XNB,XNV,CDF,CSF,CRE,
      CVD,CIN
      READ(50,*) (Q(I),I=1,6,1)
      IF (CDF .LE. 0) THEN
       CDF=0.01
        IF (XNV.LE.1) CDF=0.005
       END IF
      IF (CSF .LE. 0) CSF=0.005
      IF (CRE .LE. 0) CRE=0.0005
      IF (CVD .LE. 0) CVD=0.5
      WRITE(60,*)'PUMP PERFORMANCE - John Tuzson &
      Associates.1995.'
      WRITE(60,*)
      WRITE(60,*)TITLE
      WRITE(60,*)
      WRITE(60,*)'  Q      DN     D1     DS     D2     D3     B      B3
     1   D4    AREA'
      WRITE(60,11) Qd,DN,D1,DS,D2,D3,B,B3,D4,AREA
  11  FORMAT (2F7.1,8F7.3)
      WRITE(60,*)' BETA1    BETA2    BETA3    NB     NV
      DISK    SKIN    RECI
     1RC CVD     CIN'
      WRITE(60,12)
BEB1D,BEB2D,BEB3D,XNB,XNV,CDF,CSF,CRE,CVD,CIN
  12  FORMAT (3F7.2,2F7.1,1X,F7.4,F7.4,F7.4,F7.4,F7.4)
      WRITE(60.*)
      WRITE(70,*) '    Q     DH    EFF    POWER    DHTH
      WORK'
      DO 20 I=1,6,1

C Total head DHTH
      BEB2=0.0174533*BEB2D
      U2=(2*3.1416*DN/60)*D2/24
      WM2=(231*Q(I)/60)/(3.1416*D2*B*12)
      W2=WM2/COS(BEB2)

C Slip per Wiesner,F.J.:"A Review of Slip Factors for
  Centrifugal
```

```
C Impellers." ASME Paper #63B - 1966, WA/FE-18
C In slanted exit impellers, if the impeller O.D. is
  very close to
C the housing, slip is higher. Assume a smaller
  impeller: 0.97xD2.
      SIGMA=1-((COS(BEB2))**0.5/XNB**0.7)
      CT2=U2*SIGMA-WM2*TAN(BEB2)
      DHTH=U2*CT2/32.2

C Leakage. Bilgen et al.:"Leakage and Friction
  Characteristics
C of Turbulent Helical Flow in Fine Clearance." ASME
  Jr.Fluids Eng.
C Dec.1973, pp. 493 - 497
C The pressure drop is assumed to result from the
  difference between
C the head and a pressure rise outside the impeller due
  to a
C rotational speed one half of the impeller rotation.
C Assume L = 0.5, c = 0.01 (10/1000 on the diameter),
  Coeff. = 0.8
C Front seal only. No balancing flow.
      QL=0.8*3.1416*0.005*D1*12*8*(SQRT(DHTH-U2*U2/
      128.8))*0.25974
      IF ((DHTH-U2*U2/128.8).LE.0) QL=0

C Disk friction per Nece,R.E.,Daily,J.W.:"Roughness
  Effects on
C Frictional Resistance of Enclosed Rotating Disks."
C ASME Jr.Basic Eng. Sept.1960, pp.553 - 562
      DFC=1
      PDK=DFC*(CDF/2)*(62.4/32.2)*((DN*6.28/
      60)**3)*((D2/24)**5)
      DFH=PDK/(Q(I)*0.1390)
      DFEFF-DFH/DHTH

C Recirculation loss per Tuzson,J.:"Inlet Recirculation
  in
C Centrifugal Pumps." ASME 1983 Boston WA/FE Symposium
  on Performance
C Characteristics of Hydraulic Turbines and Pumps. Ed.
  P.Cooper
      IF (Q(I).GT.QD) THEN
      DRECH=0
      ELSE
```

```
      DRECH=(CRE/8)*((DN*6.28/60)**3)*(D1*D1-DS*DS)*((1-
      (Q(I)/QD))
     1**2.5)/Q(I)
      END IF

C Inlet incidence per Cornell,W.G.:"The Stall
  Performance of
C Cascades." Proc.4th Nat.Congr.Applied Mechanics,ASME
  1962,pp.1291-99
C and a sudden expansion loss. Prerotation proportional
  to C1.

      C1=(231*Q(I)/60)/(12*(D1*D1-DS*DS)*3.1416/4)
      U1=(2*3.1416*DN/60)*D1/24
      W1=(C1*C1+U1*U1)**0.5
      BEB1=0.0174533*BEB1D
      BE1=ATAN(U1/C1)
      WT1=C1*TAN(BEB1)

      BES1=2*BE1-BEB1
      RATIO1=(C)S(BE1))/(COS(BES1))
      TERM1=SQRT((RATIO1)**2-((COS(BEB1))/(COS(BES1))))
      IF (RATIO1.LT.0) TERM1=-TERM1
      XLAM1=RATIO1-TERM1
      DQIN12=(W1*W1/64.4)*(((1-((XLAM1*COS(BE1))/
      COS(BEB1)))/XLAM1)
     1**2)

C Inlet loss assumes that half of tangential velocity
  head difference
C is lost

C     DQIN12=0.5*((U1*U1)-(WT1*WT1))/64.4
C     IF (DQIN12.LT.0) DQIN12=-DQIN12

C Impeller skin friction loss DQSF12 based on pipe
  friction.
      DHYD12=(B*D2*3.1416*COS (BEB2))/
      (B*XNB+D2*3.1416*COS(BEB2))
      DQSF12=CSF*((D2-D1)/(COS(BEB2)*DHYD12))*(W2+W1)*
      (W2+W1)/257.6

C Impeller diffusion loss DQDIF: If inlet to exit
  relative velocity
```

274 APPENDIX

```
C ratio exceeds 1.4 then part of the vel.head
  difference is lost.
      DQDI F=0
      DIF=W1/W2
      IF (DIF.GT.1.4) THEN
      DQDIF=0.25*(DIF*DIF-2)*W2*W2/64.4
      END IF

C Calculates diffuser inlet conditions.
      CT3=CT2*D2/D3
      CM3=(231*Q(I)/60)/(12*3.1416*D3*B3)
      C3=(CT3*CT3+CM3*CM3)**05
      BEB3=0.0174533*BEB3D
      CQ3=(231*Q(I)/60)/(12*COS(BEB3)*3.1416*D3*B3)
      DQIN23=CIN*(C3*C3-CQ3*CQ3)/64.4
      IF ((C3-CQ3).LT.0) DQIN23=0
      BE3=ATAN (CT3/CM3)

C Diffuser skin friction loss DQSF34
      DHYD34=(B3*D3*3.1416*COS(BEB3))/
      (B3*XNV+D3*3.1416*COS(BEB3))
      DQSF34=CSF*((D3-D1)/(COS(BEB3)*DHYD34))*(CQ3+C1)*
      (CQ3+C1)/257.6
C Diffuser expansion loss DQVD
C Assume CVD of CQ3 velocity head is lost plus a
  diffusion loss
      IF ((CQ3/C1).GT.1.4) THEN
      DQVD=((CVD+0.25)*CQ3*CQ3-0.5*C1*C1)/64.4
      ELSE
      DQVD=CVD*CQ3*CQ3/64.4
      END IF

C For SINGLE STAGE, VOLUTE pumps set XNV=1. Then:
      IF (XNV.LE.1) THEN
      CQ3= (231*Q(I)/60)/(AREA*12)
      DQIN23=CIN*(C3*C3~CQ3*CQ3)/64.4
      IF ((C3-CQ3).LE.0) DQIN23=0
      DQSF23=CSF*(3.1416*D3/SQRT(AREA/3.1416))*(CQ3*CQ3/
      64.4)
      DQSF34=CSF*(D3/2*SQRT (AREA/3.1416))*(CQ3*CQ3)/
      64.4
      C4=(231*Q(I)/60)/(12*3.1416*D4*D4/4)
      DQVD=CVD*((CQ3*CQ3) - (C4*C4))/64.4
      END IF
      IF (DQVD.LT.0) DQVD=0
```

```
      IF (DQIN23.LT.0) DQIN23=0

C Actual head DH.
      DH=DHTH-(DQIN12+DQSF12+DQSF23+DQIN23+DQDIF+DQSF34+
     DQVD)
C The effect of leakage flow QL is taken into account:
      EFF=(DH/(DHTH+DFH+DRECH)*100)*(Q(I)/(Q(I)+QL))
      WORK=(DH/EFF)*100
      POWER=DH*Q(I)/(2.3*1714)
      BE1D=BE1*180/3.1416
      BE3D=BE3*180/3.1416

      WRITE(60,*) '   Q(I)    DH     EFF    DHTH    DQIN12
     1DQSF12 DQDIF'
      WRITE(60,10) Q(I),DH,EFF,DHTH,DQIN12,DQSF12,DQDIF
  10  FORMAT (8X,3F7.2,8X,4F7.2)
      WRITE(60,*)' DQIN23 DQSF23 DQSF34 DQVD   DFH
     DRECH    QL
     1    U1      U2'
      WRITE(60,15) DQIN23,DQSF23,DQSF34,DQVD,DFH,DRECH,
     QL,U1,U2
  15  FORMAT (1X,6F7.2,7X,3F7.2)
      WRITE(60,*) ' C1    W1     WT1    W2     WM2      CT2
     CT3     C3
     1 CQ3    C4'
      WRITE(60, 13) C1,W1,WT1,W2,WM2,CT2,CT3,C3,CQ3,C4
  13  FORMAT (1X,10F7.2)
      WRITE(60,*)
      WRITE(70,14) Q(I),DH,EFF,POWER,DHTH,WORK
  14  FORMAT (1X,1F9.1,4F7.2)
  20  CONTINUE
      CLOSE(UNIT=50)
      CLOSE(UNIT=60)
      CLOSE(UNIT=70)
      END
```

APPENDIX

```
C     John Tuzson & Associates, Evanston,IL. 1/29/99
C       The INLET code calculates approximate meridional
        velocities
C     at the inlet of a centrifugal pump, since the
      solution of
C     three-dimensional, partial differential equations
      is difficult.
C     The 20 points of the meridional hub and shroud
      contours are
C     input:RR,ZR,RT,ZT. The radii of curvature on hub
      and shroud
C     RHOR, RHOT, location of the leading edge
      RRL,ZRL,RTL,ZTL, the
C     Flow Rate Q gpm, need to be specified.
C      A line RMIN,ZMIN, is defined from the leading edge
      on the
C     hub perpendicular to the shroud line, on which J=9
      points are
C     located. Starting with a velocity of one at the
      hub, successive
C     velocities are calculated using an incremental
      factor of
C     (1+(distance between successive points / radius of
      curvature))
C     Radii of curvature are linearly interpolated
      between the values
C     on the hub and shroud. Finally the velocities are
      scaled to
C     give the flow rate Q when integrating across the
      inlet.

      DIMENSION RT(20),ZT(20),RR(20),ZR(20),RMIN(9),
      ZMIN(9),WM(9),B(20)

      OPEN (UNIT=2 0, FILE='INLET. DAT',STATUS='OLD')
C     OPEN (UNIT=30, FILE='INLET.OUT',STATUS='UNKNOWN')

C     Edit file MIXED.DTA by renaming INLET.DAT, deleting
      THER
C     and THET and adding the last data line.
      READ(20,*) (RR(I), 1=1,20,1)
      READ(20,*) (ZR(I), 1=1,20,1)
      READ(20,*) (RT(I), 1=1,20,1)
      READ(20,*) (ZT(I), 1=1,20,1)
      READ(20,*) RRL,ZRL,RTL,ZTL,RHOH,RHOS,Q
```

```
      WRITE (20,*)
      WRITE(20,*) 'John Tuzson and Associates,
      Evanston,IL 1/28/99'
      WRITE (20,*)
      WRITE(20,*) ' INLET'
      WRITE (20,*)

C   Calculates the perpendicular to the shroud line
    from RRL,ZRL
      DO 100 I=1,20,1
      B(I)=SQRT((RT(I)-RRL)**2+(ZT(I)-ZRL)**2)
 100  CONTINUE
      DO 104 I=1,18,1
      CRIT=(B(I+2)-B(I+1))*(B(I+1)-B(I))
      IF (CRIT.LT.0) THEN
      BMIN=B(I+1)
      RMIN(9)=RT(I+1)
      ZMIN(9)=ZT(I+1)
      RMIN(1)=RRL
      RMIN(1)=ZRL
      END IF
 104  CONTINUE

C   Angle with the radial direction
      ALFA=ATAN((ZMIN(1)-ZMIN(9))/(RMIN(9)-RMIN(1)))
      WM(1)=1.0

C   Coordinates on the line,radius of curvature,
    velocity.
      DO 101 J=2,9,1

      RMIN (J) =RMIN (J-1) + (COS (ALFA) ) *BMIN/8
      ZMIN (J) =ZMIN (J-1) - (SIN (ALFA) ) *BMIN/8
      RHO=RHOH+ (RHOS-RHOH) *J/8
      WM(J)=WM(J-1) * (1+BMIN/ (8*RHO))
 101  CONTINUE

C   Integrates for flow rate QS assuming WM(1)=1.0 in/s
      WM(1) =0.5
      WM(9)=WM(9)/2
      QS=0.0
      DO 102 J=1,9,1
      QS=QS+WM (J) *3.1416*RMIN (J) *BMIN/8
 102  CONTINUE
```

```
      C    Scale velocities to flow rate Q
      WM(1)=WM(1)*2*Q*(231/60)/(12*QS)
      WM(9)=WM(9)*2*Q*(231/60)/(12*QS)
      DO 103 J=2,8,1
      WM(J)=WM(J)*Q*(231/60)/(12*QS)
 103  CONTINUE

      C     WRITE(30,*) ' INLET'
C     WRITE(30,*) ' John Tuzson & Associates,
      Evanston,IL 1999'
      WRITE(20,*)
      WRITE(20,*) ' Meridional Velocity Distribution
      from Hub to Shroud'
      WRITE(20,*)
      WRITE(20,10) (WM(J), J=1,9,1)
  10  FORMAT (1X,9F8.3) -
      WRITE(20, *)
      WRITE(20,*) ' RRL ZRL RTL ZTL RHOH RHOS
     1 Qgpm'
      WRITE(20,*)
      WRITE(20,11) RRL,ZRL,RTL,ZTL,RHOH,RHOS,Q
  11  FORMAT (1X,6F9.3,1X,1F9.1)
      CLOSE (UNIT=20)
C     CLOSE (UNIT=30)
      END
```

APPENDIX **279**

```
C Program IMPEL


C John Tuzson & Associates, Evanston, IL -2/1/99
C Calculates blade coordinates of mixed flow pumps with
  blades
C that produce the same angular momentum for
  corresponding points
C on hub, mean and shroud. ALFA in the meridional plane,
  is the
C angle from the radial to the tangent to mean
  meridional streamline.
C N RPM =0 is diffuser or return passage. XN is exponent
  of
C sinusoidal angular momentum distribution:
C CT(I)=RM(21)*CT(21)*((0.5*(1~COS(I/20)))**XN)/R(I)
C The meridional velocity WM changes linearly from WM1
  to WM21
C The angular momenta: AMT,AMM,AMR are calculated and
  printed out.
C The slip SIGMA increases linerarly from inlet to
  exit.
C The blade angles BETAT,BETAM,BETAR (actual, not
  projected)
C are printed out at the end

      DIMENSION
RT(21),ZT(21),THET(21),RM(21),ZM(21),THEM(21),RR(21)
     1,ZR(21),THER(21),ALFA(21),B(21),RC(21),ZC(21),
      THEC(21),AMT(21)
     2,AMM(21),AMR(21),BETAT(21),BETAM(21),BETAR(21)
      OPEN(UNIT=10, FILE='IMPEL.DAT',STATUS='OLD')
      OPEN(UNIT=20, FILE='IMPEL.OUT',STATUS='UNKNOWN')
      OPEN(UNIT=30, FILE='IMPEL.GRF',STATUS='UNKNOWN')
      OPEN(UNIT=40, FILE='IMPEL.DTA',STATUS='UNKNOWN')
      READ(10,*)  (RM(I),I=1,21,1)
      READ(10,*)  (ZM(I),I=1,21,1)
      READ(10,*)  Q,N,T,XNB,WM1,WM21,CT21,XN,SIGMA21

C Smoothing of input
      DO 101 I=1,19,1
      RM(I+1)=(RM(I)+2*RM(I+1)+RM(I+2))/4
      ZM(I+1)=(ZM(I)+2*ZM(I+1)+ZM(I+2))/4
  101 CONTINUE
```

APPENDIX

```
      WRITE(20,*) 'John Tuzson & Associates, Evanston,IL
2/1/99'
      WRITE(20, *)
      WRITE(20,*) 'Mixed flow impeller blade
      coordinates.'
      WRITE(20,*) 'Same energy addition on all
      steamlines.'
      WRITE(20, *)
      WRITE(20,*) ' RT     ZT     THET    RM     ZM    THEM
     1RR     ZR     THER'

C Calculates velocities on main streamline.
      DO 100 1=1,20,1
      ALFA(I) =1.5708-ATAN C (RM (1+1)-RM (I))/(ZM(1+1)
      -ZM(I)))
      DM=SQRT((RM(I+1)-RM(I))**2+(ZM(I+1)-ZM(I))**2)
C Linear angular momentum CT*R distribution from inlet
  to exit:
C       WTM=(6.28/60)*N*RM(I)/12-CT21*RM(21)*I/
      (RM(I)*20) - Abandoned.
C Slip distribution
      SIGMA=1-(1-SIGMA21)*I/20
C Relative tangential velocity with sinusoidal angular
  momentum
C distribution
      IF (N.LE.0) THEN
      WTM=RM21*CT21*((0.5*(1-C05(3.1416*1/20)))**XN)/
      RM(I+1)
      ELSE
      WTM=(6.28/60)*N*RM(I+1)*SIGMA/12-
      CT21*RM(21)*((0.5*(1-COS(3.1416*
      1I/20)))**XN)/RM(I+1)
      END IF


C Meridional velocity
      WM=WM1-(WM1-WM21)*I/20
      ARCM=(WTM/WM)*DM
C Wrap angle
      THEM(1)=0
      THEM(I+1)=THEM(I)+(ARCM/RM(I+1))
C Passage width with vane blockage:
      BETAM(I)=ATAN(WTM/WM)

100 CONTINUE
```

APPENDIX

```
C Shroud and hub coordinates in meridional plane
      DO 102 I=1,20,1
      ZT(I)=ZM(I)-0.5*B(I)*COS(ALFA(I))
      RT(I)=RM(I)+0.5*B(I)*SIN(ALFA(I))
      ZR(I)=ZM(I)+0.5*B(I)*COS(ALFA(I))
 102  CONTINUE

C Circumferential angle wrap angles
      DO 105 I=1,20,1
      SIGMA=1-(1-SIGMA21)*I/20
      DT=SQRT((RT(I+1)-RT(I))**2+(ZT(1+1)-ZT(I))**2)
      IF (N.LE.0) THEN
      WTT=RM(21)*CT21*((0.5*(1-COS(3.1416*I/20)))**XN)/
      RT(I+1)
      ELSE
      WTT=(6.28/60)*N*RT(I+1)*SIGMA/12-
      CT21*RM(21)*((0.5*(1-COS))
     1(3.1416*1/20)))**XN)/RT(I+1)
      END IF
      WM=WM1-(WM1-WM21)*1/20
      ARCT=(WTT/WM)*DT
      THET(1)=0
      THET(I+1)=THET(I)+(ARCT/RT(I+1))

      DR=SQRT((RR(I+1)-RR(I))**2+(ZR(I+1)-ZR(I))**2)
      IF (N.LE.0) THEN
      WTR=RM(21)*CT21*((0.5*(1-C0S(3.1416*1/20)))**XN)/
      RR(I+1)
      ELSE
      WTR=(6.28/60)*N*RR(I+1)*SIGMA/12-
      CT21*RM(21)*((0.5*(1-COS
     1(3.1416*1/20)))**XN)/RR(I+1)
      END IF
      WM=WM1-(WM1-WM21)*1/20
      ARCR=(WTR/WM)*DR
      THER(1)=0
      THER(I+1)=THER(I)+(ARCR/RR(I+1))
 105  CONTINUE

C Calculates reference circle for plotting and
      converts angles
      WRITE(30,*) ' rt    zt    thet    thet2    rr    zr
     1ther    ther2    rc    zc    thec'
      DO 103 I=1,20,1
      THEC(1)=0
```

282 APPENDIX

```
      RC(I)=RT(20)*1.1
      ZC(I)=ZT(10)
      THEC(I+1)=THEC(I)+(360/20)
      THET(I)=THET(I)*(180/3.1416)
      THET2=THET(I)+360/XNB

      THEM(I)=THEM(I)*(180/3.1416)
      SMT=RT(I)*1.414
      THER(I)=THER(I)*(180/3.1416)
      THER2=THER(I)+360/XNB
      SMR=RR(I)*1.414
      WRITE(20,12)
RT(I),ZT(I),THET(I),RM(I),ZM(I),THEM(I),RR(I)
    1,ZR(I)  ,THER(I)
      WRITE(30,12)
RT(I),ZT(I),THET(I),THET2,RR(I),ZR(I),THER(I),THER2,
    1RC (I),ZC(I),THEC(I),SMT,SMR
   12 FORMAT (13F8.3)
  103 CONTINUE

C Calculates the angular momentum on tip AMT and root
  AMR sqft/sec

      WRITE(20,*)
      WRITE(20,*) '    Angular Momentum    Blade Angles'
      WRITE(20,*)
      DO 104 1=1,20,1
      SIGMA=1-(1-SIGMA21)*1/20
      WM=WM1-(WM1-WM21)*1/20
      DT=SQRT((RT(I+1)-RT(I))**2+(ZT(I+1)-ZT(I))**2)
      CTT=((6.24/60)*N*RT(I)*SIGMA/12)-
      WM*RT(I)*(THET(I+1)-THET(I))*
     1(3.1416/180)/DT
      AMT(I)=CTT*RT(I)/12
      DM=SQRT((RM(I+1)-RM(I))**2+(ZM(I+1)-zM(I))**2)
      CTM=((6.24/60)*N*RM(I)*SIGMA/12)-
      WM*RM(I)*(THEM(I+l)-THEM(I))*
     1(3.1416/180)/DM
      AMM(I)=CTM*RM(I)/12
      DR=SQRT((RR(I+l)-RR(I))**2+(ZR(I+1)-ZR(I))**2)
      CTR=((6.24/60)*N*RR(I)*SIGMA/12)-
      WM*RR(I)*(THER(I+l)-THER(I))*
     1(3.1416/180)/DR
      AMR(I)=CTR*RR(I)/12
```

```
      BETAT(I)=(180/3.1416)*ATAN((THET(I+1)-
      THET(I))*RT(I+1)*
    1(3.1416/180)/DT)
      BETAM(I)=(180/3.1416)*ATAN((THEM(I+1)-
      THEM(I))*RM(I+1)*
    1(3.1416/180)/DM)
      BETAR(I)=(180/3.1416)*ATAN((THER(I+1)-
      THER(I))*RR(I+1)*
    1(3.1416/180)/DR)

      IF (I.LE.19) WRITE(20,13) AMT(I),AMM(I),AMR(I),
      BETAT(I),
    1 BETAM(I),BETAR(I)
  13 FORMAT (6F9.3)
 104 CONTINUE

      WRITE(40,*) (RR(I),I=1,20,1)
      WRITE(40,*) (ZR(I),I=1,20,1)
      WRITE(40,*) (THER(I),I=1,20,1)
      WRITE(40,*) (RT(I),I=1,20,1)
      WRITE(40,*) (ZT(I),I=1,20,1)
      WRITE(40,*) (THET(I),I=1,20,1)

      CLOSE (UNIT=10)
      CLOSE (UNIT=20)
      CLOSE (UNIT=30)
      CLOSE (UNIT=40)

      END
```

284 APPENDIX

```
C Program MIXED

C John Tuzson & Associates, Evanston, IL -12/29/97,8/
  29/96,4/18/95
C Calculates blade coordinates of mixed flow pumps with
  blades
C that can be pulled from casting molds in direction
  GAMAREF (angle
C from axial) and THEO (angle circumferencially). ALFA
  is angle from
C radial in meridional plane and tangent to mean
  meridional streamline.
C ALFA(10)=GAMAREF and THEM(10)=THEO. N RPM =0 is
  diffuser.
C XN is exponent of sinusoidal angular momentum
  distribution on the
C mean streamline.
C The meridional velocity WM changes linearly from WM1
  to WM21
C The angular momenta: AMT,AMM,AMR are calculated and
  printed out.
C The slip SIGMA increases linerarly from inlet to
  exit.

C The blade angles BETAT,BETAM,BETAR (actual, not
  projected)
C are printed out at the end

      DIMENSION
     RT(21),ZT(21),THET(21),RM(21),ZM(21),THEM(21),RR(21)
     1,ZR(21),THER(21),ALFA(21),B(21),RC(21),ZC(21),
      THEC(21),AMT(21)
     2,AMM(21),AMR(21),BETAT (21),BETAM (21),BETAR(21)
      OPEN(UNIT=10,FILE='MIXED.DAT',STATUS='OLD')
      OPEN(UN IT=20,FILE='MIXED.OUT',STATUS='UNKNOWN')
      OPEN(UNIT=30,FILE='MIXED.GRF',STATUS='UNKNOWN')
      OPEN(UNIT=40,FILE='MIXED.DTA',STATUS='UNKNOWN')
      READ(10,*)  (RM(I),I=1,21,1)
      READ(10,*)  (ZM(I),I=1,21,1)
      READ(10,*)  Q,N,T,XNB,WM1,WM21,CT21,GAMAREF,XN,
      SIGMA21

C Smoothing of input
      DO 101 I=1,19,1
      RM(I+1)=(RM(I)+2*RM(I+1)+RM(I+2))/4
```

```
      ZM(I+1)=(ZM(I)+2*ZM(I+1)+ZM(I+2))/4
101 CONTINUE

      WRITE(20,*) 'John Tuzson & Associates, Evanston,IL
      8/25/99'
      WRITE(20,*)
      WRITE(20,*) 'Mixed flow impeller blade coordinates'
      WRITE(20,*)
      WRITE(20,*) '     RT     ZT     THET    RM     ZM
      THEM
     1RR  ZR    THER'

C Calculates wrap of main streamline, and relative
  velocities.
      DO 100 I=1,20,1
      ALFA(I+1)=1.5708-ATAN((RM(I+1)-RM(I))/(ZM(I+1)-
      ZM(I)))
      DM=SQRT((RM(I+1)-RM(I))**2+(ZM(1+1)-ZM(I))**2)
C Uniform angular momentum CT*R distribution from inlet
  to exit:
C     WTM=(6.28/60)*N*RM(I)/12-CT21*RM(21)*I/
      (RM(I)*20)-Abandoned.
C Slip
      SIGMA=1-(1-SIGMA21)*I/20
C Sinusoidal angular momentum distribution on mean
  streamline
      IF (N.LE.0) THEN
      WTM=CT21*((0.5*(1-COS(3.1416*1/20)))**XN)
      ELSE
      WTM=(6.28/60)*N*RM(I+1)*SIGMA/12-CT21*RM(21)*0.5*
      (1-COS(3.1416*
     11/20))/RM(Iö1)
      END IF

      WM=WM1-(WM1-WM21)*I/20

      ARCM=(WTM/WM)*DM
      THEM(1)=0
      THEM(I+1)=THEM(I)+(ARCM/RM(I+1))
C With vane blockage:
      BETAM(I)=ATAN(WTM/WM)
      B(I)=((Q*231/60)/(WM*12))/((RM(I)*6.28)-(T*XNB/
      COS(BETAM(I))))
100 CONTINUE
```

```
C      IF (GAMAREF.LE.0) GANAREF=ALFA(10)*180/3.1416
C Calculates tip and root coordinates

       GAMAREF=GAMAREF*3.1416/180
       DO 102 I=1,20,1
       ZT(I)=ZM(I)-0.5*B(I)*COS(ALFA(I))-
       0.5*B(I)*(TAN(ALFA(I)-
      1GAMAREF)) *SIN(ALFA(I))
       RT(I)=RM(I)+0.5*B(I)*SIN(ALFA(I))-
       0.5*B(I)*(TAN(ALFA(I)-
      1GAMAREF))*COS(ALFA (I))
        THEO=THEM (10)
        PT=2*RM(I)*(COS(THEM(I))*COS(THEO)+SIN(THEM(I))*
        SIN(THEO))
        QT=RM(I)*RM(I)-RT(I)*RT(I)
        ST=(-PT/2)+SQRT((PT*PT/4)-QT)
        THET(I)=ATAN( (RM(I)*SIN(THEM(I))+ST*SIN(THEO))/
        (RM(I)*COS(THEM
      1(I))+ST*COS(THEO))
        IF (THET(I).LT.0) THET(I)=THET(I)+3.1416

       ZR(I)=ZM(I)+0.5*B(I)*COS(ALFA(I))+0.5*B(I)*
       (TAN(ALFA(I)-
      1GAMAREF))*SIN(ALFA(I))
       RR(I)=RM(I)-0.5*B(I)*SIN(ALFA(I))+0.5*B(I)*
       (TAN(ALFA(I)-
      1GAMAREF))*COS(ALFA(I))
       PR=-2*RM(I)*(COS(THEM(I))*COS(THEO)+SIN(THEM(I))*
       SIN(THEO))
       QR=RM(I)*RM(I)-RR(I)*RR(I)
       IF ((PR*PR/4-QR).LT.0) THEN
C        RR(I)=0
C        ZR(I)=0
       THER(I)=0
       ELSE
       SR=(-PR/2)-SQRT((PR*PR/4)-QR)
       THER(I)=ATAN( (RM(I)*SIN(THEM(I))-SR*SIN(THEO))/
       (RM(I)*COS(THEM
      1(I))-SR*COS(THEO))
       END IF
       IF (THER(I).LT.-0.7854) THER(I)=THER(I)+3.1416
  102 CONTINUE

C Calculates reference circle for plotting and converts
  angles
```

```
      WRITE(30,*) ' rt    zt    thet    thet2    rr    zr
     1ther    ther2    rc    zc    thec'
      DO 103 I=1,20,1
      THEC (1)=0
      RC(I)=RT(20)*1.1
      ZC(I)=ZT(10)
      THEC(I+1)=THEC(I)+(360/20)
      THET(I)=THET(I)*(180/3.1416)
      THET2=THET(I)+360/XNB
      THEM(I)=THEM(I)*(180/3.1416)
      SMT=RT(I)*1.414
      THER(I)=THER(I)*(180/3.1416)
      THER2=THER(I)+360/XNB

      SMR=RR(I)*1.414
      WRITE(20,12) RT(I),ZT(I),THET(I),RM(I),ZM(I),
      THEM(I),RR(I)
     1,ZR(I),THER(I)
      WRITE(30,12) RT(I),ZT(I),THET(I),THET2,RR(I),
      ZR(I),THER(I),THER2
     1RC(I),ZC(I),THEC(I),SMT,SMR
   12 FORMAT (13F8.3)
  103 CONTINUE

C Calculates the angular momentum on tip AMT and root
AMR sqft/sec

      WRITE(20,*)
      WRITE(20,*) '      Angular Momentum      Blade Angles'
      WRITE(20,*)
      DO 104 I=1,20,1
      SIGMA=1-(1-SIGMA21)*1/20
      WM=WM1-(WM1-WM21)*1/20
      DT=SQRT((RT(I+1)-RT(I))**2+(ZT(I+1)-ZT(I))**2)
      CTT=((6.24/60)*N*RT(I)*SIGMA/12)-WM*RT(I)*
      (THET(I+1)-THET(I))*
     1(3.1416/180)/DT
      AMT(I)=CTT*RT(I)/12
      DM=SQRT((RM(I+1)-RM(I))**2+(ZM(I+1)-ZM(I))**2)
      CTM=((6.24/60)*N*RM(I)*SIGMA/12)-
      WM*RM(I)*(THEM(I+1)-THEM(I))*
     1(3.1416/180)/DM
      AMM(I)=CTM*RM(I)/12
      DR=SQRT((RR(I+1)-RR(I))**2+(ZR(I+1)-ZR(I))**2)
```

```
      CTR-((6.24/60)*N*RR(I)*SIGMA/12)-
      WM*RR(I)*(THER(I+1>THER(I))*
     1(3.1416/180)/DR
      AMR(I)=CTR*RR(I)  /12

      BETAT(I)=(180/3.1416)*ATAN((THET(I+1)-
      THET(I))*RT(I+1)*
     1(3.1416/180)/DT)
      BETAM(I)=(180/3.1416)*ATAN((THEM(I+1)-
      THEM(I))*RM(I+1)*
     1(3.1416/180)/DM)
      BETAR(I)=(180/3.1416)*ATAN((THER(I+1)-
      THER(I))*RR(I+1)*
     1(3.1416/180)/DR)

      IF (I.LE.19) WRITE(20,13) AMT(I),AMM(I),AMR(I),
      BETAT(I),
     1 BETAM(I),BETAR(I)
  13  FORMAT (6F9.3)
 104  CONTINUE

      WRITE(20,*)
      THEO=THEO*180/3.1416
      GAMAREF=GAMAREF*180/3.1416
      WRITE(20,*) THEO,GAMAREF

      WRITE(40,*)  (RR(I),I=1,20,1)
      WRITE(40,*)  (ZR(I),I=1,20,1)
      WRITE(40,*)  (THER(I),I=1,20,1)
      WRITE(40,*)  (RT(I),I=1,20,1)
      WRITE(40,*)  (ZT(I),I=1,20,1)
      WRITE(40,*)  (THET(I),I=1,20,1)

      CLOSE (UNIT=10)
      CLOSE (UNIT=20)
      CLOSE (UNIT=30)
      CLOSE (UNIT=40)

      END
```

```
C      John Tuzson & Associates, Evanston, IL. 8/25/99
C      CUTS is a program for preparing cuts through an
       impeller blade
C      at equally spaced planes perpendicular to the axial
       direction.
C      The hub and tip (shroud) lines are input by the
       coordinates
C      R,Z,Theta. The origine of z is at the impeller
       inlet and is
C      counted positive into the eye. The blades are
       defined by
C      straight lines between the corresponding points on
       hub and tip.
       DIMENSION RT(20),ZT(20),THET(20),RR(20),ZR(20),
       THER(20),
      2Z (20, 6) , R (20, 6) ,THE (20, 6)
       OPEN (UNIT=55, FILE='CUTS. DAT', STATUS='OLD')
       OPEN (UNIT=65, FILE= ' CUTS . OUT', STATUS='
       UNKNOWN')
       READ(55,*) (RR(I),I=1,20,1)
       READ(55,*) (ZR(I),I=1,20,1)
       READ(55,*) (THER(I),I=1,20,1)
       READ(55,*) (RT(I),I=1,20,1)
       READ(55,*) (ZT(I),I=1,20,1)
       READ(55,*) (THET(I),I=1,20,1)

       DO 100 1=1,20,1
       THET (I) =THET (I) *3.1416/180
       THER(I)=THER(I) *31416/180
  100  CONTINUE
C      Calculation of the cuts. Coordinates set to 0
       outside the blade.
       DO 101 J=1,6,1
       DO 102 1=1,20,1
       Z(I,J)=ZT(1)=(ZR(20)-ZT(1))*(J-1(/5
       R(I,J)-RR(I)+(RT(I)-RR(I)*RT(I)*(Z(I,J)-ZR(I))/
       (ZT(I)-ZR(I))
       X=RR(I)*RT(I)*((COS(THET(I)))*(SIN(THER(I)))-
       ((COS (THER(I)))
      1*SIN(THET(I))))/(RT(I)*(SIN(THET(I)))RR(I)*
       SIN(THER(I)))
       ALFA=ATAN((RT(I)*SIN(THET(I)))/
       (X+RT(I)*COS(THET(I))))
       THE(I,J)=ALFA+ASIN((X*SIN(ALFA)) /R(I,J))
```

```
C The coordinates are set to zero outside the blade
        IF (R(I,J).GT.RT(I)) THEN
        R(I,J)=0
        Z(I,J)=0
        THE(I,J)=0
        END IF
        IF (R(I,J).LE.RR(I)) THEN
        R(I,J)=0
        Z(I,J)=0
        THE (I,J)=0
        END IF
        THE (I, J) =THE (I, J) *180/3.1416
  102 CONTINUE
  101 CONTINUE
        WRITE(65,*) ' Coordinates r, z, theta of the six
        cuts'
        WRITE (65,*)
        DO 103 I=1,20,1
        WRITE(65,10) R(I,1),Z(I,1),THE(I,1),R(I,2),
        Z(I,2),THE(I,2),
        1R(I,3),Z(I,3),THE(I,3),R(I,4),Z(I,4),
        THE(I,4),R(I,5),Z(I,5)
       2THE(I,5),R(I,6),Z(I,6),THE(I,6)
   10   FORMAT (18F9.3)
  103   CONTINUE
        CLOSE (UNIT=55)
        CLOSE (UNIT=65)
        END
```

```
C    John Tuzson & Associates, Evanston, IL. 12/10/98

C    Program BLADE2.FOR calculates the radii and centers
     of the arcs of
C    circle which make up radial impeller blades.

     DIMENSION R(5),BETA(5),RHO(5),THETA(5),XO(5),
     YO(5),X(5),Y(5)
     OPEN (UNIT=30,FILE='BLADE.OUT',STATUS='UNKNOWN')
     OPEN (UNIT=20,FILE='BLADE.DAT',STATUS='OLD')
C    number of stations=5, R(I)=radii, BETA(I)=blade
     angles from
C    radial, RHO=radii of curvature, XO(I),YO(I)
     coordinates
C    of center of curvature, THETA(I)=blade coordinate
     from X axis.
     READ(20,*) BETA(1),BETA(5)
     READ(20,*)  (R(I), I=1,5,1)
     THETA(1)=0.0
     WRITE(30,*)    'John Tuzson & Associates,
     Evanston, IL. 12/10/98'
     WRITE(30,*)
     WRITE(30,*)    'BLADE2 - Impeller blade
     coordinate calculation'
     WRITE(30,*)
     WRITE(30,*)    'R = Radius, BETA = Blade angle
     from radial'
     WRITE(30,*)    'THETA = Angle at center of
     rotation from X axis'
     WRITE(30,*)    'RHO = Radius of curvature'
     WRITE(30,*)    'XO,YO = Coordinates of center of
     curvature'
     WRITE(30,*)
     WRITE(30,*)'   R    Beta    Theta    Rho    XO    YO'
     WRITE(30,*)
     WRITE(30,11) R(1), BETA(1), THETA(1)
11   FORMAT (F7.3,2x,F7.3,2x,F7.3)

     DO 24 I=2,5,1
     BETAY=BETA(1)
     RATIO=(R(I)-R(1)) /(R(5)-R(1))
     BETAX=(BETA(5) -BETAY) *RATIO
     BETA(I) =BETAY+BETAX
24   CONTINUE
```

```
      DO 25 I=1,4,1
      X(1)=R(1)
      Y(1)=0.0
      THETA(1) =0.0
      BETA(I)=(90-BETA(I))*3.1416/180
      BETA(I+1)=(90-BETA(I+1))*3.1416/180
      THETA(I)=THETA(I)*3.1416/180
      DEN=(R(I+1)*COS(BETA(I+1)))-(R(I)*COS(BETA(I)))
      RHO(I)=(R(I+1)*R(I+1)-R(I)*R(I))/(2*DEN)
      XO(I+1)=X(I)-RHO(I)*COS(BETA(I)-THETA(I))
      YO(I+1)=Y(I)+RHO(I)*SIN(BETA(I)-THETA(I))
      S2=R(I)*R(I)+RHO(I)*RHO(I)-
      2*R(I)*RHO(I)*COS(BETA(I))
      S=SQRT (S2)
      ALFA2=ACOS((R(I+1)*R(I+1)+S2-RHO(I)*RHO(I))/
      (2*R(I+1)*S))
      ALFA1=ATAN(XO(I+1)/YO(I+1))
      THETA(I+1)=1.5708-ALFA1-ALFA2
      IF (ALFA1.GT.0.0) THETA(I+1)=4.7124-ALFA1-ALFA2
      X(I+1)=R(I+1)*COS(THETA(I+1))
      Y(I+1)=R(I+1) *SIN(THETA(I+1)
      BETA(I+1)=(1.5708-BETA(I+1)) *180/3.1416
      THETA(I+1)=THETA(I+1) *180/3.1416

      WRITE(30,10) R(I+1),BETA(I+1),THETA(I+1),RHO(I),
      XO(I+1),YO(I+1)
10    FORMAT (F7.3,2x,F7.3,2x,F7.3,2x,F7.3,2x,F7.3,
      2x,F7.3)
25    CONTINUE
      CLOSE (20)
      CLOSE (30)
      END
```

```
C    Program EXREC. John Tuzson & Associates,Evanston,IL
     6/4/98

C   The program calculates the flow rate at the onset of
    exit
C   recirculation in mixed flow pumps. It is assumed that
    at
C   the best efficiency point the work input or
    theoretical
C   head or angular momentum R*CT is uniform across the
    impeller
C   exit. It can be shown that this condition implies
    uniform
C   meridional velocity CMO, which can be calculated
    from the
C   flow rate at best efficiency and the exit geometry.
    The
C   exit blade angle as a function of radius is then
    defined.
C   The first order differential equation in CM is
    integrated
C   starting at the hub radius, where CM=0 is the
    condition for
C   the onset of impeller exit recirculation, and ending
    at the
C   shroud radius. At each step the flow rate is
    incremented,
C   the final value corresponding to the flow rate at the
    onset
C   of recirculation.
       OPEN(UNIT=50, FILE='EXREC.DAT',STATUS='OLD')
       OPEN (UNIT=60, FILE= ' EXREC. OUT', STATUS=
       'UNKNOWN')
C   Theoretical head THT ft, Flow rate QO gpm, shroud
    radius RSH,
C   hub radius RH, exit width B, rpm XN, slip SIG,
    number of steps N
       READ(50,*) HTH,QZ,RSH,RH,B,XN,SIG,N
       WRITE(60,*) HTH,QZ,RSH,RH
       WRITE(60,*) B,XN, SIG,N
       WRITE(60,*)
       CMO=(QZ*231/60)/(3.1416*B*(RSH+RH))
       OMG=6.28*XN/60
       DA=144*HTH*32 .2/OMG
```

APPENDIX

```
        CM=0
        Q=0
        R=RH
        DR=(RSH-RH)/(N-1)
        TB1=(SIG*OMG*RH/CMO)-(DA/(RH*CMO))
        IF (TB1.LT.0.6) TB1=0.6
        BETA1=(180/3.1416)*ATAN(TB1)
        WRITE(60,*)RH, BETA1

        DO 10 I=2,N,1
        TB=(SIG*OMG*R/CMO)-(DA/(R*CMO))
        IF (TB.LT.0.6) TB=0.6
        DTB=(SIG*OMG/CMO)+(DA/(R*R*CMO))
        IF (TB.LE.0.6) DTB=0
        X=((3*SIG-1)*OMG*TB-(1-SIG)*OMG*R*DTB)*CM
        Y=(TB*TB/R+TB*DTB)*CM*CM
        Z=2* (1-SIG) *OMG*OMG*R*SIG
        DEN=(1+TB*TB) *CM+(1-SIG)*OMG*R*TB
        DCM=((X-Y+Z)/DEN)*DR
        CM=CM+DCM
        Q=Q+6.28*R*B*CM*DR/(RSH-RH)
        R=R+DR
        TB=(SIG*OMG*R/CMO) - (DA/ (R*CMO))
        BETA=(180/3.1416) *ATAN (TB)
        WRITE(60,*)R,BETA,CM
10      CONTINUE
        Q=Q*60/231
        WRITE(60,*)
        WRITE(60,*) Q
```

INDEX

Absolute velocity, 61
Aeration, 40
Air leaks, 40
Angular momentum, 14, 63, 78, 101, 124, 146, 147, 151, 154, 169, 194, 210
Atmospheric pressure, 109
Axial flow pumps, 2, 253
Axisymmetric flow, 13

Balance chamber, 85
Balance of forces, 10, 237
Balancing, 184
Balancing holes, 85
Base circle, 78, 145
Bearings, 184, 220, 254
Bernoulli's equation, 10, 67, 101
Blade
  angle, 64, 79, 147
  coordinates, 150, 158
  design, 167
  loading, 67, 69, 72, 151
  passing frequency, 46, 69, 78
  projection, 140
  thickness, 153
  wake, 34
Blade-to-blade calculation, 265
Blockage, 79, 92, 124, 137

Boundary layer assumptions, 92
Boundary layer thickness, 94
Boundary layers, 83, 91
Bubbly flow, 203

Calibration of program, 129
Casting, 149
Cavitation, 40, 80, 107, 194
  coefficient, 110
  damage, 116
  of liquids, 108
Centrifugal acceleration, 13, 227
Characteristic curve, 47
Clearance, 82
Clearance space, 99
Compressibility, 46
Computer programs, 1, 121, 264
Conservation of energy, 10, 101, 113
Conservation of mass, 9
Control volume, 9
Coriolis acceleration, 13, 103, 227
Coriolis erosion tester, 230
Cost estimate, 258
Coupling, 45
Critical speed, 183
Crude, 205
Curved diffusers, 33
Cylindrical coordinates, 13

295

Diffuser, 30, 142, 207
Diffuser loss, 127
Diffusion, 72
Diffusion loss, 126, 218
Dimensionless numbers, 7
Dimensions, 6
Discharge coefficient, 29
Disk friction, 82, 99, 127, 218
Disk pump, 83, 248
Displacement thickness, 95
Dissolved gases, 201
Double suction, 56
Drag, 23
  coefficient, 24
Drives, 257
Dynamic head, 38
Dynamic viscosity, 94

Eddy viscosity, 19, 93
Efficiency, 37, 55, 128, 134, 249
Ejector, 235, 248
  efficiency, 236
  performance, 238
Electric motors, 38, 43, 257
End suction, 55
Energy conservation, 37
Engineering societies, 3
Equal energy flow, 12
Equation of continuity, 9, 101
Erosion, 223
  material removal, 223
  mechanisms, 224
  pattern, 227
  resistance, 230
  testing, 230
Euler's equation, 63
Exit blade angle, 124, 137, 195
Exit recirculation, 189
Expeller blades, 86

Flow angle, 64, 79
Flow coefficient, 134
Fluid coupling, 212
Fluid machinery types, 250
Force balance, 101
Four quadrant, 46
Free streamline, 22, 96
Fundamental equations, 9

Head coefficient, 134
Head-flow curve, 48, 52, 129, 197, 207, 219
Hub, 141, 147
Hub-to-shroud calculation, 265
Hydraulic components, 45
Hydraulic performance, 37
Hydraulic ram, 249
Hydrocarbon fluids, 217

Impact erosion, 224
Impeller, 61
Impeller boundary layers, 103
Impeller inertia, 46
Incidence, 79, 124, 147, 191
Inducer, 117
Inertia forces, 17
Inertia load, 46
Inlet
  blade angle, 139
  diameter, 138
  flow, 79
  recirculation, 127, 189
  velocity, 139, 166
Instability, 198
Irrotational flow, 12

Jet erosion testers, 230
Jet-wake flow, 73

Kinematic viscosity, 94

Leading edge, 63, 78, 113, 139, 149, 156
Leading edge vortex, 96, 194
Leakage, 83, 86, 127
Lift, 23
Liquid-gas mixtures, 201
Load curve, 48
Loss coefficients, 122
Losses, 249

Manufacturing methods, 149
Maximum packing, 225
Mechanical design, 147
Meridional velocity, 61
Mixing loss, 34
Modeling, 264

INDEX **297**

Momentum, 10
Multistage pumps, 76

Natural frequency, 184
Navier–Stokes equations, 9, 266
Nomenclature, 123
Normalized head, 204
Normalized torque, 204
NPSHA, 40, 81
NPSHR, 40, 80, 109, 139
Nuclear reactor, 203
Number of blades, 138
Number of vanes, 143
Numerical examples, 161

Off-design, 38, 53, 69, 77, 189
Oil wells, 205
Operating point, 47
Orifices, 29

Partial admission, 252
Particle concentration, 226
Particles, 223
Performance, 51, 203
Performance calculation, 121, 133, 147, 163, 217
Phase separation, 203, 210
Pipes, 27
Pitot tube pump, 252
Positive displacement pump, 250
Potential flow, 12, 265
Power demand, 41, 52, 129
Prerotation, 124, 137
Pressure
  coefficient, 7, 30
  exchange, 34, 75
  head, 11
  pulsations, 46, 76
  recovery, 30, 76
  side, 67, 69
Pressure loss in pipes, 28
Price, 259
Product development, 248
Pump configuration, 55
Pump design procedure, 133
Pump load, 46
Pump speed, 38
Pump-out blades, 86

Recirculation, 189
Reference pressure, 5
Regenerative pump, 252
Relative velocity, 61
Return flow passage, 76, 142, 158, 175
Reynold's number, 17, 93, 218
Rotating axes, 14
Rotating disc, 98
Rothalpy, 67, 191, 210
Rotor dynamics, 147, 183

Scaling, 55
Sealing sleeve, 85
Seals, 184, 253
Selection, 37, 53
Self priming, 78
Separation, 20, 70, 72, 79, 96, 167, 189
Separation point, 22
Separation streamline, 191
Service factor, 38
Shaft deflection, 185
Shaft seal, 254
Shear, 82, 91
  flows, 13
  force, 17
  layer, 22, 73, 96
  pump, 83, 248
Shroud, 141, 147
Shut off head, 41
Side channel pump, 252
Similarity, 53
Single stage, 55
Skin friction, 98, 126, 218
Sliding erosion, 225
Slip, 129, 151
Slip coefficient, 65, 124, 137
Slurry pots, 230
Slurry pumps, 86, 104, 223
Specific energy, 225
Specific speed, 54, 112, 249
Specifications, 37
Speed changes, 46
Speed selection, 134, 250, 257
Stability, 19, 48, 96, 194
Stable separation, 75
Stagnation point, 23, 93
Stagnation pressure, 11, 12
Standard pipe sizes, 38

Standards, 81, 109, 146, 184
Start up, 46
Static head, 11, 38
Suction side, 69
Suction specific speed, 111
Sudden expansion, 31, 75, 79, 124
Surface tension, 202
Surging, 49, 198
Swirling flow, 13
System modeling, 41

Technology development, 247
Theoretical head, 61, 68, 123, 134
Tip speed, 63
Tip vortex, 79
Torque converter, 213
Total head, 38
Total pressure, 11
Trailing edge, 63, 156
Trailing vortices, 23, 75, 104, 228
Transient operation, 46
Trust bearing, 85
Turbine, 249, 253
Turbulence, 19, 75, 92, 103, 266
Turbulent mixing, 203
Two-dimensional blade, 153, 173
Two-phase flow, 107, 201

Unbalance, 184

Vacuum, 108
Vaned diffuser, 75
Vaneless diffusers, 100
Vaneless space, 76
Vapor pressure, 82, 109
Variable frequency drive, 257
Velocity head, 11
Velocity triangle, 64, 136
Vertical pump, 57
Vibration, 117
Viscosity, 7, 18, 91, 217
Viscous forces, 17
Viscous pumps, 220
Void fraction, 202
Volute, 77, 126, 144, 177
Volute tongue, 77, 145
Vortex flow, 13, 20

Wake, 24
Wake mixing loss, 75
Wall friction, 92, 98, 126
Water hammer, 46
Wear, 224
Wear ring, 85, 86, 187, 220, 253
Work input, 129
Wrap, 152, 171